Key Concepts STEM

SCIENCE TECHNOLOGY ENGINEERING MATH

PHYSICS

ELECTRICITY AND ELECTRONICS

General Editor: Anne O'Daly

BROWN BEAR BOOKS

Published by Brown Bear Books Ltd
4877 N. Circulo Bujia, Tucson, AZ 85718
USA
and
Studio G14, Regent Studios, 1 Thane Villas,
London N7 7PH, UK

Cataloging-in-Publication data available upon request.

ISBN 978-1-83572-078-3 (ALB)
ISBN 978-1-83572-083-7 (paperback)
ISBN 978-1-83572-088-2 (ebook)

Children's Publisher: Anne O'Daly
Design Manager: Keith Davis
Picture Manager: Sophie Mortimer
Consultant: Chris Woodford

Manufactured in the United States of America
CPSIA compliance information: Batch#AG/5665

Picture Credits
Front Cover: Shutterstock: chuyuss.
Interior: iStock: scanrail 28-29, Thinkstock 14; **NASA**: CXC/CFA/S. Wolk et al 44-45; **Public Domain**: Morcheeba 46, Reinraum 10-11, Stahlkocher 49; **Shutterstock**: Cagla Acikgoz 21, Andersphoto 45, Arogant 39, Alexey Avdeev 34-35, Muhammad Bin jamal 32, Eimantas Buzas 58-59, chuyuss 1, 27, Doug McLean 40, mehmetcan 50-51, Morphart Creation 18, raigvi 54, Frederico Roslagno 42-43, Sashkin 56, Vladimir3d 22.

STEM

Key Concepts in STEM: Physics describes the key theories in the study of physics and their practical applications. This book revolves around the electron, the tiny negatively-charged particle that is found in the atoms of all elements. It also looks at the development of electronics, from early vacuum tubes and transistors through to the discovery of microchips and semiconductor materials. Explanatory diagrams and informative photographs, detailed features on related aspects of the topics covered and the main scientists involved in the advancement of physics, and definitions of key "Science Words" all enhance the coverage. "Try This" sidebars outline experiments that learners can follow as a first step to practical investigations.

Publisher's note to educators and parents: Our editors have carefully reviewed the web sites included at the end of this book to ensure that they are suitable for students. Many web sites change frequently, however, and we cannot guarantee that a site's future contents will continue to meet our high standards of quality and educational value. Be advised that students should be closely supervised whenever they access the Internet.

Contents

The Empty Atom 4
The Elusive Electron 6
Static Electricity 8
Attraction and Repulsion 10
Detecting Charge 14
Electric Fields 18
Charged Atoms 20
Ions and Electrolysis 22
Moving Charges 26
Potential Difference 28
Current and Stored Charge 32
Resistance and Power 34
Direct and Alternating Current 38
Vacuum Tubes and Their Uses 40
X-Rays and Beyond 44
Semiconductor Materials 46
Solid-State Devices 50
Making Chips 54
Electronic Memories 58
Glossary 62
Further Research 63
Index 64

THE EMPTY ATOM

The whole of modern electronics is built on the properties of one tiny particle that is found in the atoms of all elements—the electron.

At the end of the 19th century most physicists were convinced that chemical elements existed in the form of atoms—tiny units of matter that are normally undivided and indivisible. But nothing at all was known about their internal structure.

The first glimpse into the atom came when the English physicist William Crookes (1832–1919) invented the Crookes tube. It was a glass bottle in which the air pressure could be reduced at least a millionth of its normal value. Two electrodes protruded into the low-pressure gas. When a high voltage was applied across the electrodes, glowing colored patches and bands of light appeared in the tube. The light changed in complex ways as the pressure and voltage were altered. Crookes was able to show that the bands were caused by something moving from the negative electrode (the cathode) toward the positive electrode (the anode). He gave the name "cathode rays" to these streams of mysterious objects.

Another English physicist, J. J. Thomson (1856–1940), applied electric and magnetic fields to cathode rays. He found that the rays consisted of identical, negatively charged particles. They were the same no matter what sort of gas was in the tube, and they seemed to be much lighter than even the lightest atom. Thomson claimed that these particles were all fragments of atoms. They were soon named "electrons" from the Greek word

ELECTRONS

In an atom the negatively charged electrons surround the positively charged nucleus. The regions in which they are found are known as shells. Shown here is an atom of carbon, with its six electrons.

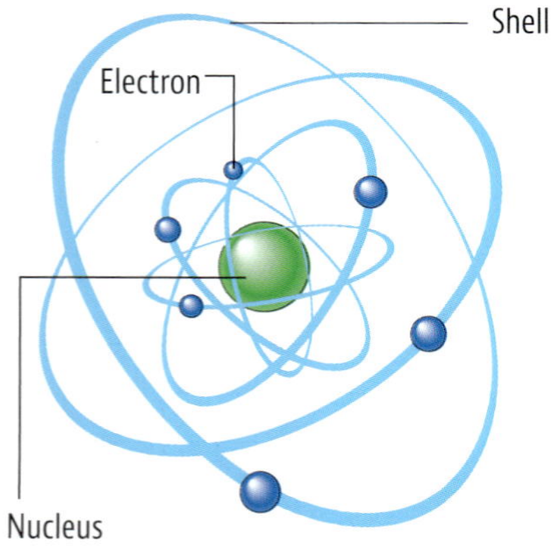

THE HEART OF THE ATOM

Ernest Rutherford (1871-1937) allowed positively charged alpha particles to bombard a thin foil of gold. Most passed through, but a small number were deflected through large angles; they had bounced off a positively charged core, or nucleus, within the atom.

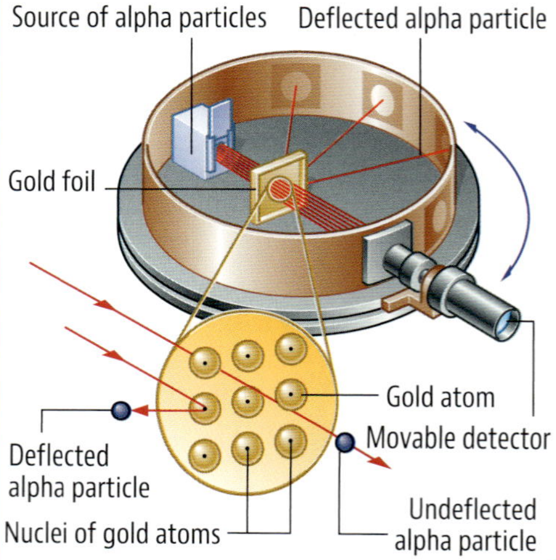

SCIENCE WORDS

Anode: A positive electrical terminal on a device such as an electron tube.
Atom: The smallest part of a chemical element that can exist on its own. It has a central nucleus, surrounded by electrons.
Cathode: A negative electrical terminal on a device such as an electron tube.

elektron, which means "amber" (static electricity was first made by rubbing amber).

Cathode rays consist of electrons that have emerged at the cathode, and flow from the cathode toward the anode. They collide with gas atoms on the way, knocking further electrons out of those atoms.

If the normally electrically neutral atom contains negatively charged electrons, it had to contain equal amounts of balancing positive charge. So how were the electrons and the positive charge arranged in the atom?

Deeper Into the Atom

A New Zealand-born physicist, Ernest Rutherford, probed deeper into the atom in 1911. For this he used alpha particles, which are given out in radioactivity. They are helium atoms that have lost their electrons and are thus positively charged. Rutherford allowed them to strike a thin sheet of gold. Most went right through, but a tiny number were deflected, some of them quite strongly. The only way Rutherford could explain this was to suppose that the positive charge was concentrated in a tiny core, or nucleus, at the center of the atom. Most alpha particles missed the nuclei, but some came close, were repelled, and bounced back. The nucleus proved to have one ten-thousandth the diameter of the atom. Electrons roam through the outer parts of the atom—relatively speaking, a truly enormous volume.

TRY THIS

The Scale of the Atom

Electronics involves electrons, and electrons come from atoms. An atom consists of a central nucleus surrounded by one or more electrons. In this project you will make a scale model of a simple atom.

What to do

The simplest atom is the hydrogen atom. It has a nucleus consisting of a single proton, and a single electron orbits this nucleus. In this model, a tennis ball represents the nucleus and a pea represents the electron. Their sizes are in about the right proportion, so you can see how much smaller the electron is. But how big is the hydrogen atom?

Go outdoors to a tennis court (or an area of similar size). Put the tennis ball in one corner of the court and the pea in the corner diagonally opposite. If you don't have access to a tennis court, place the pea about 27 yards (25 meters) from the tennis ball. This distance is in about the right proportion to the "proton" and the "electron." You can see that a hydrogen atom is mostly empty space, as are all other kinds of atom. Keep hold of the tennis ball and ask a friend to hold the pea and run around you in a big circle about 55 yards (nearly 50 meters) across. This is what the electron does in a hydrogen atom.

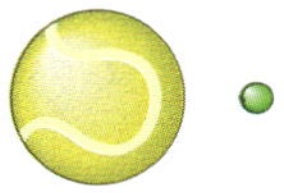

This model of the hydrogen atom is roughly to scale. It demonstrates that an atom consists mostly of empty space.

THE ELUSIVE ELECTRON

When electrons were identified, ingenious experiments revealed their properties. Electrons proved to be responsible for the most significant properties of all matter. When they break loose from their atoms, electrons can flow as electric currents.

William Crookes carried out experiments with the Crookes tube that showed cathode rays consist of something moving from the cathode to the anode. He was able to demonstrate that these objects could exert pressure on obstacles placed in their path. J. J. Thomson had shown that the objects were probably very small.

The American physicist Robert A. Millikan (1868–1953), who measured the charge on the electron in 1909, confirmed these facts. When combined with earlier measurements by Thomson, Millikan showed that the mass of the electron was approximately one two-thousandth of the mass of a hydrogen atom.

Physicists now understand what electric currents are. They consist of electrons that

WILLIAM CROOKES

By the time William Crookes died in 1919, his greatest contribution to science had been used to create crude images, and had earned him a knighthood in 1897. Within just a few years of his death, his invention became the basis of a new method of communication: television. The tube was also used as one of the earliest sources of x-rays.

Crookes also invented improved methods of making sugar from beet, dyeing textiles, and extracting silver and gold from their ores. He also publicized the benefits of electric lighting. In 1861, Crookes discovered the metal thallium from the new and then unknown pattern of colors in its spectrum, and he went on to study it.

THE CROOKES TUBE

An air pump reduces the pressure in the glass tube. When a high voltage is applied between the two electrodes (cathode and anode), negatively charged particles move from the cathode (negative electrode) toward the anode (positive electrode). These "cathode rays" overshoot and strike the far end of the tube, making it glow. They cast a shadow if an obstacle is put in their path. In his experiment Crookes used a thin piece of tin in the shape of a Maltese cross.

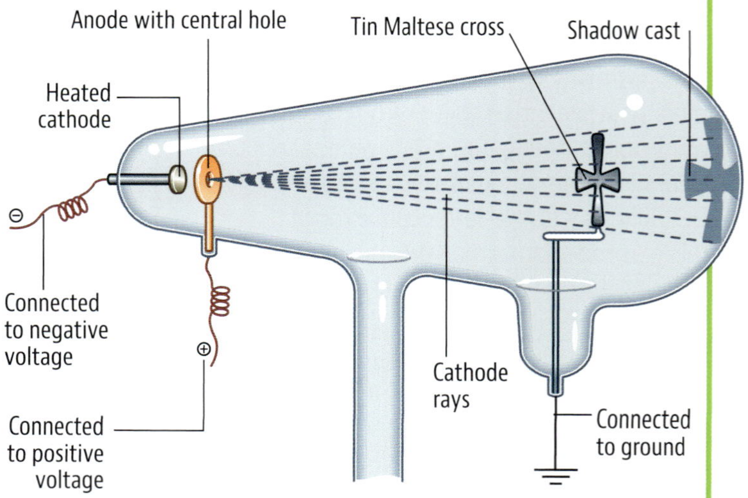

have broken free from their atoms and are drifting through, say, a metal wire, or sometimes through space. Currents generate heat as the electrons bump into atoms. If they generate enough heat, the temperature of the material rises until it glows—this is what happens in the tungsten filament in an electric light bulb.

Structure of the Atom

The experiments by Ernest Rutherford (see page 5) had shown that the outer regions of atoms were occupied by electrons, while at the center was the nucleus. It was natural for scientists to think of the atom as being like a miniature Solar System, with the electrons playing the role of planets traveling round the central nucleus, which stood for the Sun.

The electrons were held in place not by gravity, but by the attraction between their negative charges and the positive charge on the nucleus. However, there was a problem with this image. According to the theories that existed, electrons moving around within the atom like this should give out all their energy in a brief burst of electromagnetic radiation as they spiraled into the nucleus. All atoms should collapse within a fraction of a second.

PRESSURE OF ELECTRONS

In one of the demonstrations that can be carried out with a Crookes tube, a vane that is free to rotate is placed inside it. When a voltage is applied, the vane is pushed around until it is parallel with the stream of cathode rays. This experiment suggests that cathode rays are particles rather than waves.

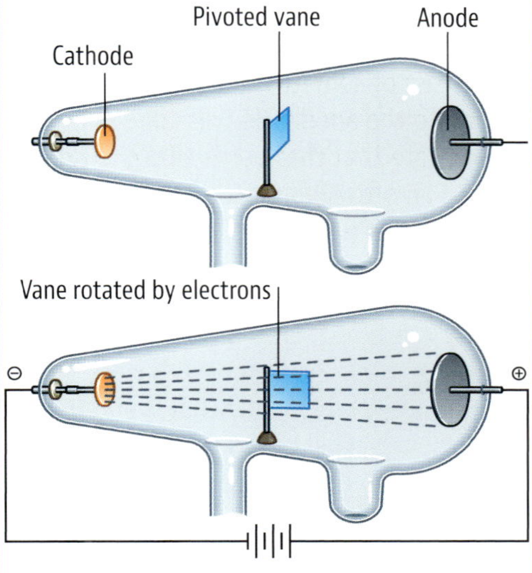

SCIENCE WORDS

Crookes tube: An early experimental vacuum tube that generated cathode rays.
Electromagnetic radiation: Energy that is transmitted in the form of electromagnetic waves.
Electron: A subatomic particle, found in every atom, which carries negative charge.
Quantum theory: A theory based on the idea that light is emitted in separate packets, or quanta (which are also known as photons).

In 1912, the Danish physicist Niels Bohr (1885–1962) suggested that electrons occupy only certain orbits, each having its own definite amount of energy. The only way to move between orbits was to make an abrupt jump from one to another, giving out or taking in energy in the form of electromagnetic radiation in amounts corresponding to the differences in the energy levels of the orbits. Atoms do not collapse because their lowest energy levels are filled with electrons. From this picture of the atom, scientists gradually developed quantum theory, which is the basis of modern physics.

STATIC ELECTRICITY

Electricity is one of the essentials of modern living. It provides light, heat, communications, and power. Even ancient peoples knew about electricity. This was the type of electricity called static electricity, which is created when an electric charge builds up on an object.

The word "static" means stationary, and static electricity is concerned with stationary electric charges. More than 2,500 years ago, an ancient Greek scientist named Thales of Miletus (c. 624–c. 546 BCE) discovered that when a piece of amber is rubbed with a cloth, it can then be used to pick up small pieces of paper, rather like a magnet picks up pins. When 17th-century scientists began to study this effect, they made up the word "electricity" from the Greek word *elektron*, which means "amber." But what is static electricity, and where does it come from? Like so many questions in physics, the answer lies with atoms.

STEALING ELECTRONS

Static electricity is created from contact between two materials that do not conduct electricity, but which can generate it when rubbed together. Rubbing a plastic ruler with a cloth gives the ruler a charge of positive electricity. The charge arises because electrons from the ruler are transferred to the cloth. The cloth becomes negatively charged.

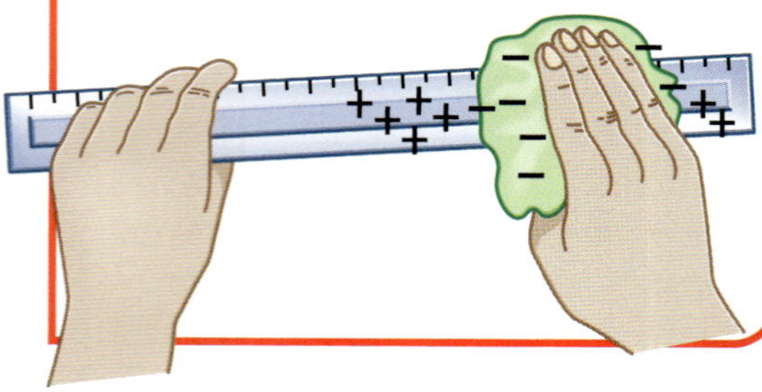

SCIENCE WORDS

Charge: A property of some subatomic particles and some objects that makes them exert forces on one another. There are two types of charge: negative charge and positive charge.

Nucleus: (plural: nuclei) The central part of an atom, which consists of protons and neutrons.

Static electricity: An electric charge on an object that has lost or gained electrons.

Atomic Electricity

An atom consists of a central nucleus that is surrounded by one or more electrons. The nucleus has a positive charge to balance the negative charges of the electrons. Overall, the atom has no charge. If you rub a plastic ruler with a cloth, some of the electrons in the atoms of the ruler "rub off" onto the atoms in the cloth. As a result, the plastic ruler gains a positive charge (because it is less negative); the cloth takes on a negative charge (because it has gained extra electrons).

Combing your hair about 20 times with a plastic comb gives the comb a negative charge. The positively charged hair may then stand on end, and may crackle.

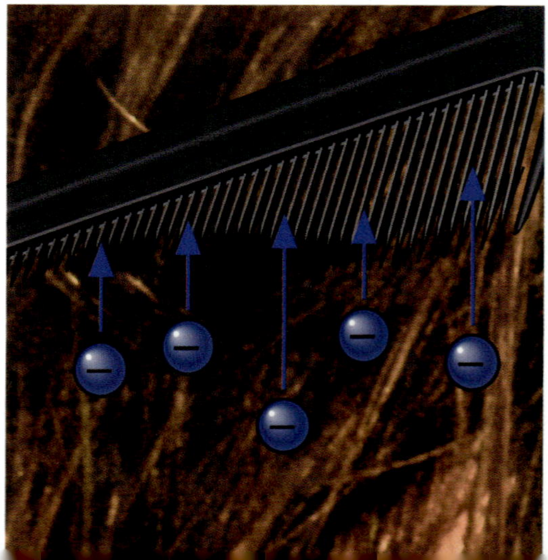

Making Sparks

The charging is achieved by rubbing. You will probably have come across static electricity. If you walk across a nylon carpet, you might receive a slight electric shock afterward when you touch a metal object such as a radiator or handrail. If, on a dry day, you take off a sweater by pulling it over your head, you can sometimes hear the crackle of tiny electric sparks. In a darkened room, you can even see the sparks. And if you rub a balloon vigorously on your clothes, it will become electrically charged and can be stuck to a wall. A charged balloon can also make somebody's hair stand on end.

Kinds of Charge

Whether the charge is negative or positive depends on the materials being rubbed together. A scientific theory in the 2010s suggested that some statically charged objects contain pockets of both negative and positive charges, and that it is the total percentage of charge of each object that leads to the attraction of the objects. In general, rubbing moves charges—electrons—from one material to the other. As a result, each material acquires a charge. The material that gains extra electrons acquires a negative charge. The material with fewer electrons becomes positively charged. It has been seen that rubbing a cloth on plastic produces a positive charge on the plastic. Rubbing a glass rod with a piece of silk also produces a positively charged glass rod. If a hard rubber rod is rubbed with fur, the rod acquires a negative charge.

It is important to remember that in none of these cases is electricity made. The electricity is already there in the atoms of the materials. Rubbing moves charges (electrons) from one material to another, resulting in each material acquiring a charge, either negative or positive.

TRY THIS

Attractive Plastic

Static electricity is electricity that does not move. It consists of stationary electric charges. With a few everyday items you can investigate some of the properties of electrically charged objects.

What to do

Take a piece of tissue paper about 10 in x 3 in (25 cm x 7.5 cm). Using scissors, cut the paper into long narrow strips, but don't cut right to the edge of the paper—leave each strip attached (see below). Hold the paper with one hand. With your other hand, move a comb through your hair quickly several times, then bring the comb up to (but not touching) the ends of the tissue strips. They will move toward the comb.

Passing the comb through your hair charged it with negative static electricity. When you held the comb near the paper streamers, the comb made them take on a positive charge. Because unlike charges attract each other, the paper was pulled toward the comb.

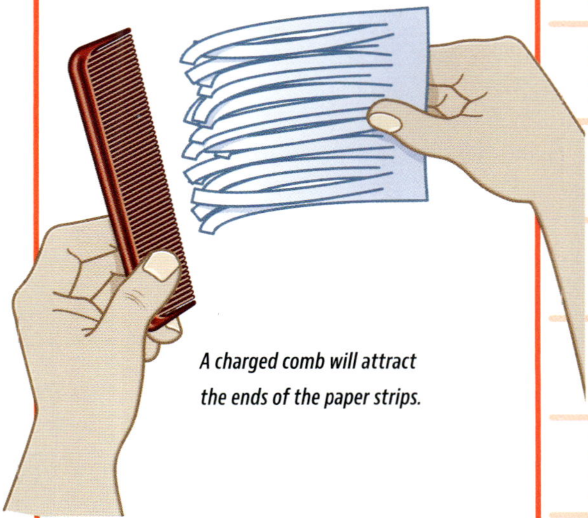

A charged comb will attract the ends of the paper strips.

ATTRACTION AND REPULSION

A Van de Graaff generator is a machine that creates static electricity. When the metal sphere on top starts charging, it transfers the charge to the person who is touching it. Their hair stands on end because the hair follicles are charged in the same way and try to repel each other.

There are two kinds of static electricity—positive and negative. They can both be studied by charging objects and observing how they behave. One thing will soon become obvious: some charges attract each other, but others repel.

Rubbing can be used to give an electric charge to objects such as plastic rulers and combs. One charged object, such as a ruler, can be used to charge another object. Imagine suspending a small ball of plastic, such as Styrofoam, from the end of a thread. If a plastic ruler is then rubbed with a cloth—acquiring a positive charge—and held against the plastic ball, the ruler passes its charge onto the ball, giving it a positive charge. In a similar way, a rubber rod rubbed on fur acquires a negative charge. A small plastic ball suspended by a thread can be given a negative charge by touching it with the charged rubber rod.

In both cases, as soon as the ball becomes charged, it will swing away from the object that charged it. What is happening is that the positively charged ball is repelled—or forced away—by the positively charged ruler, and the negatively charged ball is repelled by the negatively charged rubber rod.

Similar and Different Charges

Attraction and repulsion between charges can be demonstrated by another simple experiment. Imagine suspending two small Styrofoam balls near each other. If one is given a negative charge and the other is given a positive charge, what do you think will happen? The result is shown at the top of the diagram on the facing page: the two balls swing closer together. This experiment demonstrates that unlike electric charges attract each other.

Now imagine giving both balls the same charge, a positive charge. This time the balls swing apart, demonstrating that two positive charges repel each other. If the experiment is repeated giving both balls a negative charge, the result will be the same. The two balls swing apart, demonstrating that two negative charges also repel each other. In fact, two similar charges (two positives or two negatives) always repel each other. This

can be summarized in a simple statement: unlike charges attract; like charges repel.

These experiments confirm there are two kinds of static electricity—positive and negative. They also begin to explain how a charged comb can pick up pieces of paper. Being pulled through somebody's hair gives the comb a negative charge. When the comb is placed next to the pieces of paper, it repels some of the negatively charged electrons in the paper (because like charges repel). This leaves a positive charge on the paper nearest the comb. Unlike charges attract each other, so the comb attracts the paper and picks it up.

Several practical devices make use of the forces of attraction and repulsion between charges, for example electrostatic paint sprayers and photocopying machines.

ATTRACT OR REPEL?

Two balls given different electric charges (**a**) attract each other. Two balls, both with positive charges (**b**) or both with negative charges (**c**), repel each other.
Conclusion: unlike charges attract;
like charges repel.

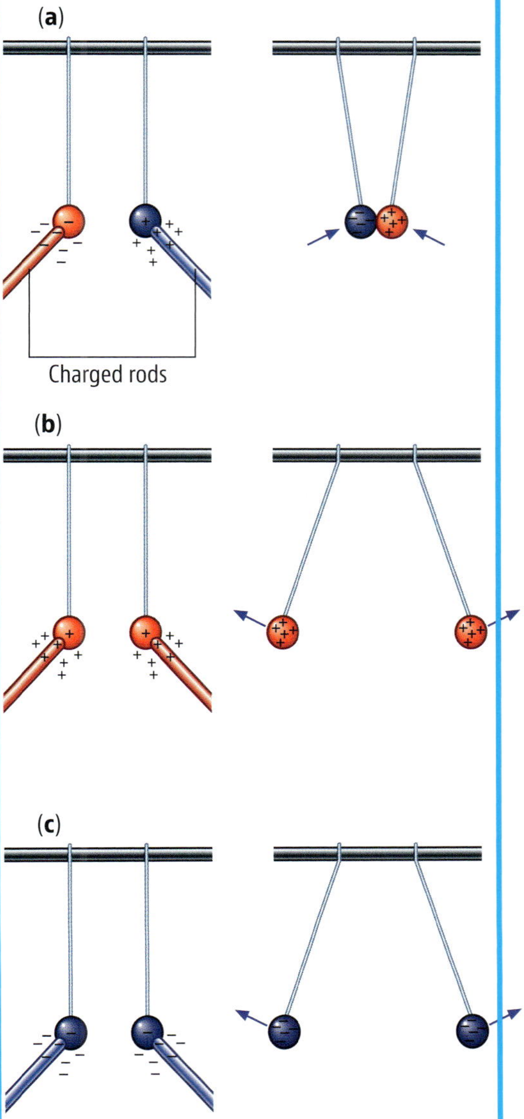

The Force That Moves

It is a basic fact of physics that the only way to make something move (or to stop something from moving) is to use force. For example, the force of gravity makes an apple fall off a tree and onto the ground. So what is the force that makes charged objects move toward or away from each other? It must be an electric force.

It is possible to repeat the three experiments illustrated on page 11, this time trying the effect of using larger or smaller charges. It is also possible to hang the charged balls closer together or farther apart. These experiments were carried out as long ago as 1785 by the French physicist Charles-Augustin de Coulomb (1736–1806). He found that the strength of the force between electric charges depends on the product of the charges (the size of one charge multiplied by the size of the other charge). He also found that the force is stronger when the charges are closer together. Mathematically, the strength of the force between charges increases as the square of the distance between them (the distance multiplied by itself) decreases.

The two relationships are combined in Coulomb's law, which can be written as:

$$F = k \frac{Q_1 \times Q_2}{d^2}$$

where F is the force, Q_1 and Q_2 are the charges, and d is the distance between them. The term k is a mathematical constant that depends on the medium—whether the charges are in air, in a vacuum, and so on. This famous equation is an example of what is called an inverse square law (because the force is proportional to the inverse of the square of the distance). There are several other inverse square laws in physics, such as the law of gravitation.

SCIENCE WORDS

Conservation of charge: In an isolated system the overall electric charge remains constant.

coulomb (C): The SI unit of electric charge.

Coulomb's Law: The force between two electric charges is proportional to the product of the charges and inversely proportional to the square of the distance between them. In other words, the force is stronger for bigger charges and weaker as the charges are moved apart.

COULOMB'S LAW

According to Coulomb's law, the larger the charges, the greater is the force between them **(a)**. Also, the strength of the force depends on the distance between the charges **(b)**. The equation **(c)** states the law.

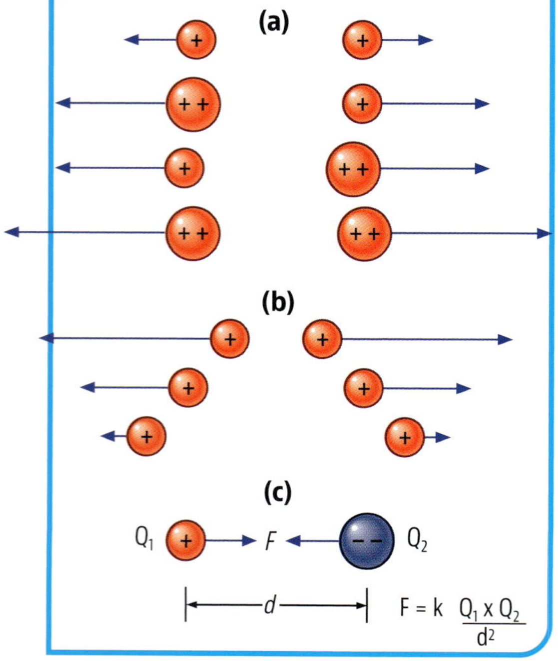

Strength and Direction

Like all forces, the force between electric charges is a vector quantity. This means that it has a certain value (its strength) and a particular direction. In the experiments so far described, the force acts in a direction along the line joining the two charges. However, if there are several charges, working out the directions becomes more complicated.

In addition to having a law named after him, Coulomb also gave his name to the unit of electric charge. It is an extremely large unit. In fact, 1 coulomb equals the charge on 60 million million million electrons (the number 6 followed by 19 zeroes). The force between two charges of 1 coulomb each situated about 3.3 ft (1 m) apart is about 2 billion pounds (9 billion newtons). In practice, the largest charges produced are no bigger than a tiny fraction of a coulomb.

When charges are very small and at "ordinary" distances apart, Coulomb's law tells us that the force of attraction or repulsion between them is also very small. That is why a charged comb will pick up only very small pieces of paper. But if the charges are extremely close together (if the distance d is very small in the equation for Coulomb's law), the quantity d^2 gets very small indeed. This fact becomes extremely important within the structure of atoms. The atomic nucleus has a positive charge, and the electrons have a negative charge. And because the nucleus and electrons are extremely close together, the force of attraction between them is very large indeed. In fact, it is such electric forces that hold atoms together. These forces are also significant in the structures of molecules, and they are responsible for keeping ions in their regular places in the internal structure of crystals.

We have already seen how to charge an object by rubbing it. The rubbing action moves some electrons either to or from the object, giving it a negative or positive charge. Separating charges or bringing them together in this way does not affect their size. The overall net charge remains the same. In fact, in an isolated system the net electric charge always remains constant. In scientific terms, this is known as the principle of conservation of charge.

TRY THIS

A Love and Hate Relationship

This project shows how like charges repel each other. Like charges are either two negative charges or two positive ones.

What to do

Blow up two balloons and tie their necks. Use a marker pen to label them – and + (the symbols for negative and positive). Tie a length of string to each balloon and tape the ends of the strings to the frame of a doorway so that the balloons hang apart from each other, as in the illustration below.

Stroke one of the balloons about 20 times on your hair or clothing. Carefully let it go, and see what happens. The stroking gave the balloon a negative electric charge that caused a positive charge to form on the other balloon. They attract each other and move together like long-lost friends.

Hold both balloons for a while so that they lose their charge. Stroke one of them again on your hair or clothing. This time, ask a friend to stroke the other balloon on your hair or clothes. Let go of the balloons. What happens now? Suddenly, they drift to opposite sides of the doorway. This is because both balloons have the same (negative) charge, and like charges repel each other. You will have to cross out the + sign on one balloon and change it to a – sign.

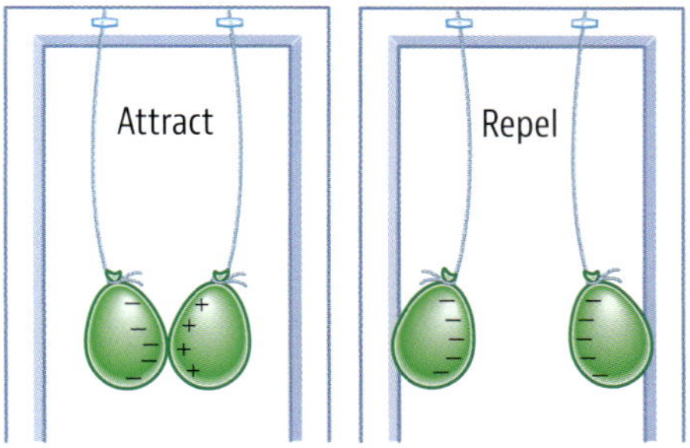

DETECTING CHARGE

A nonconducting material—an electrical insulator—can be charged by rubbing it, but how do we know when it is charged? And how do we tell if it has a positive charge or a negative charge? The best way is to use an instrument called an electroscope, which registers charge on a pair of gold leaves.

This apparatus was used by French scientists Pierre Curie (1857–1906) and Marie Curie (1867–1934). They used it to detect the presence of radioactivity. It is a gold-leaf electroscope fitted with a microscope to detect the tiniest movement of the gold leaf.

The gold-leaf electroscope has changed very little since the late 1780s, when the first such device came into use. In its simplest form, the electroscope consists of a metal rod with two thin pieces of gold foil attached to the lower end. There is a metal disk at the upper end. The metal rod passes through a block of rubber or plastic, which insulates the rod from the metal case of the electroscope. The easiest way to charge an electroscope is to take a charged object and

CHARGING BY INDUCTION

It is possible to charge an electroscope without making contact with the metal disk. A charged plastic rod is brought close to the disk **(a)** without touching it. The leaves diverge. If the disk is touched with a finger it allows the electrons to travel to the ground through the scientist's body, and the leaves will hang down. The positive charge remains on the disk and, when the rod is removed, spreads down to the gold leaves **(b)**, which then move apart once more.

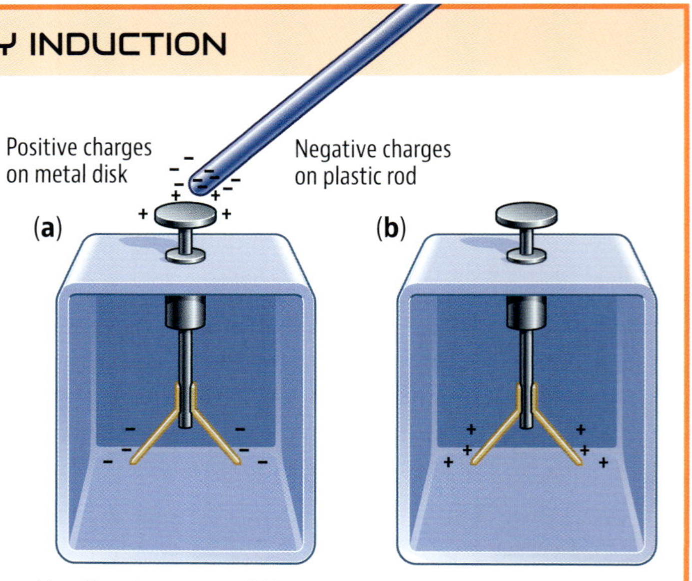

Negative charges on gold leaves, they move apart

Positive charges on gold leaves, they move apart

touch it to the metal disk. This is called charging by contact. A negatively charged object gives a negative charge to the gold leaves, and a positively charged object gives them a positive charge. In either case, the two gold leaves have the same charge as each other. Because like charges repel, the leaves diverge, or move apart. The larger the charge, the farther the leaves move apart.

Charging Without Touching

Another way of charging an electroscope is by induction. If a negatively charged object is brought near the metal disk but is not allowed to touch it, it causes electrons to be repelled down the rod. The gold leaves acquire a negative charge and they diverge.

If the disk is then touched by the scientist, electrons pass through the scientist's body to ground, and the leaves collapse, leaving the positive charge on the disk. Finally, if the source of negative charge is removed, the positive charge spreads from the disk over the rod and the leaves, which become positively charged and move apart. In this way, a negatively charged object has induced a positive charge on the leaves of the electroscope. If the test is carried out with a positively charged object, the induction method gives the leaves a negative charge.

TRY THIS

Bending Water

This simple experiment demonstrates electrostatic induction. Adjust a cold-water faucet so that the water flows as a stream of many drips moving close together. Get a large plastic spoon and rub it on your clothes to give it an electric charge. Now hold the bowl of the spoon near the water flow. The charge on the spoon induces an opposite charge on the near side of the moving drips. The unlike charges (spoon and drips) attract each other, and the stream of water is bent out of the vertical toward the spoon. The same principle is used in recycling plants to separate out the different components of trash.

TESTING CHARGES

The charge on an object can be identified by using a charged electroscope, although you have to know whether the instrument carries a negative charge or a positive charge.

The object is brought close to the disk of the electroscope. If its leaves diverge even more (**a**), the object has the same charge. If they collapse slightly (**b**), the object has the opposite charge.

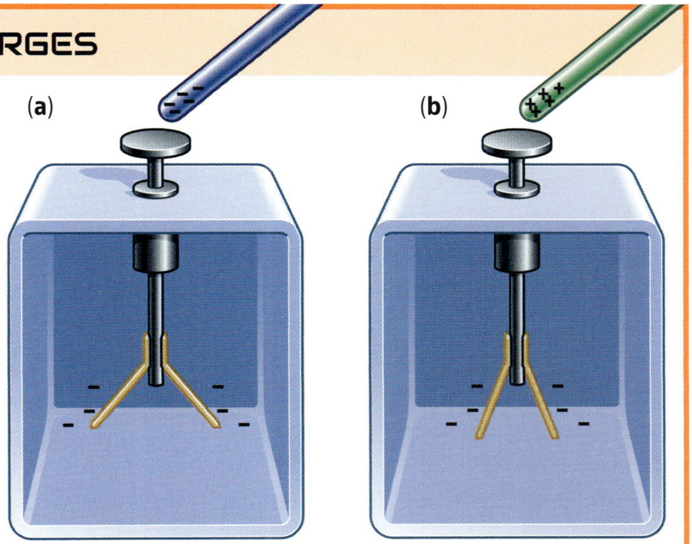

Gold leaves move farther apart — Gold leaves move closer together

TESTING FOR CONDUCTORS

A charged electroscope can be used to test whether an object is an electrical insulator or a conductor. An insulator **(a)** does not conduct away the charge, and the electroscope's leaves do not collapse. A good conductor **(b)** carries away the charge, and the leaves collapse.

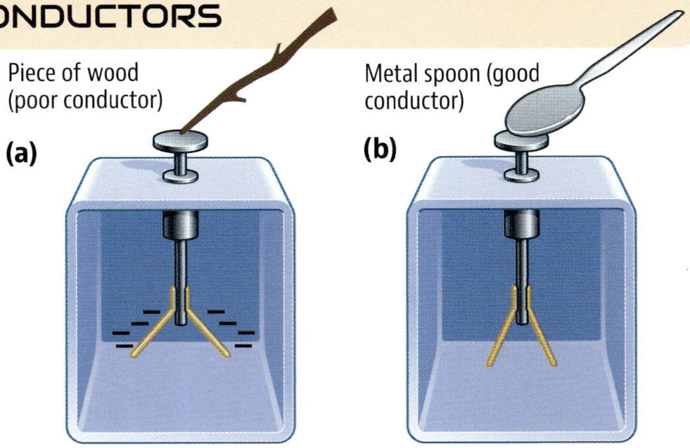

Gold leaves keep charge

Gold leaves lose charge (collapse)

Positive or Negative?

We now have a way of charging the electroscope leaves so that they are known to be either positive or negative. We can use this fact to determine the nature of the charge on another object (see sidebar page 15). Assuming that the electroscope has a negative charge on its leaves, all that needs to be done is to bring the unknown charge close to the electroscope's disk. If the leaves diverge still farther, the unknown charge is negative. If the leaves collapse, the unknown charge must be positive. However, it is better to test for a positive charge by bringing the charged object close to the disk of an electroscope with positively charged leaves. The leaves will then diverge even farther.

Conductor or Insulator?

An electroscope can provide even more information about objects, such as whether they are good or bad conductors of electricity. One way of testing an object is to touch it onto the disk of a charged electroscope. If it is a bad conductor, the charge will remain on the electroscope, and its leaves will remain apart. But if it is a good conductor, such as a metal, it will immediately conduct the charge away, and the electroscope's leaves will rapidly lose their charge and collapse.

All materials that can be given an electric charge through rubbing—glass, plastic, hard rubber, etc—are insulators. Only they can hold the charge. An attempt to charge a piece of metal by rubbing it on cloth or fur will fail. A conductor cannot normally be charged by rubbing it. Vigorously combing your hair with a metal comb will not allow you to pick up pieces of paper with the comb, as is possible with a plastic comb, which is an insulator.

SCIENCE WORDS

Conductor: A material that conducts electricity (or heat).

Electroscope: An instrument for detecting electric charge.

Electrostatic induction: The production of an electric charge in an uncharged object by a nearby charged object.

Insulator: A material that is a poor conductor of electricity, also called a nonconductor. ("Insulator" is also the name given to a poor conductor of heat.)

Distribution of Charge

With any charged object, the charge is located on its outer surface. There is no charge on the solid interior material. On a sphere, for example, the charge is evenly distributed all over the surface. The similar charges repel each other, and they move around until they are all at an equal distance from each other. On a pear-shaped object, however, there is more charge on the pointed end than on the blunt end. The sharper an object is, the more charge it will hold, and the more charge it will attract—which is why there are sometimes points on a lightning rod.

If a pointed conductor is placed near a charged object, the opposite charge is induced at the point. But this charge leaks rapidly back, leaving the opposite charge on the other end of the pointed conductor. In this way, the charge is effectively transferred from the original charged object to the blunt end of the pointed conductor. Pointed objects can be used to collect charge. For this reason, a conductor covered in dust will not hold a charge because the charge tends to leak away from the dust particles, which behave as tiny points on the surface.

For a hollow charged object the charge is always located on the outside. Over 150 years ago, the British physicist Michael Faraday (1791–1867) carried out an ingenious experiment to demonstrate this fact. He made a net bag—like a fishing net—out of a thread that conducts electricity. He then electrically charged the net and tested it inside and out. He could detect no charge inside the net, only on the outside of it. Then, pulling on a silk thread sewn to the inside corner of the net, he carefully turned it inside out. Faraday found that the electric charge was now located on the new outside of the net, and there was no charge at all remaining on what was now the new inside.

Faraday also carried out a famous experiment to investigate what is now called induction. He placed an open metal can on the disk of an electroscope (he actually used an ice bucket, and for this reason the demonstration is often called Faraday's ice-pail experiment). He lowered a charged ball deep into the can without it touching the can's sides or bottom, and the electroscope's leaves diverged. When he removed the ball, the leaves collapsed.

He repeated the sequence of steps, but this time when the leaves diverged, he touched the ball onto the inside of the can. The leaves stayed diverged; when the ball was removed, it was found to have no charge at all. It had been neutralized by the opposite charge induced on the inside of the can.

Faraday's experiment also proved that the charge on a conductor is located entirely on its outer surface.

CHARGE DISTRIBUTION

An electrically charged spherical object, shaped like an apple **(a)**, has the charge spread evenly all over its surface. With a pointed object, shaped like a pear **(b)**, the charge distribution is uneven. There is more charge around the pointed end than there is around the blunt end. A pointed conductor can even pull the charge off a charged object and neutralize it.

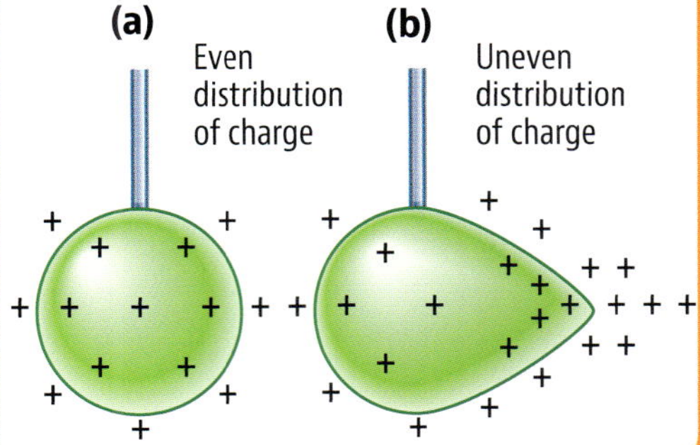

ELECTRIC FIELDS

An electric charge can influence another charge near it. For this effect to work at a distance there must be an electric field extending outward in all directions around the charge—just like the magnetic field around a magnet.

The electric field of a point charge extends in all directions. It is a force field—that is, it exerts a force on another charge placed in the field—and can therefore be represented by lines of force. The direction of the field at any particular point is, by agreement among scientists, defined as the direction in which a positive charge would move if placed at that point. So the lines of electric force move outward from a positive charge and move in toward a negative charge.

POINT CHARGE FIELDS

The lines of force around a point positive charge (**a**) move outward in all directions. The lines of force between two positive charges (**b**) diverge and curve. The same happens with two negative charges—like charges repel. There is a neutral point between the charges with no field.

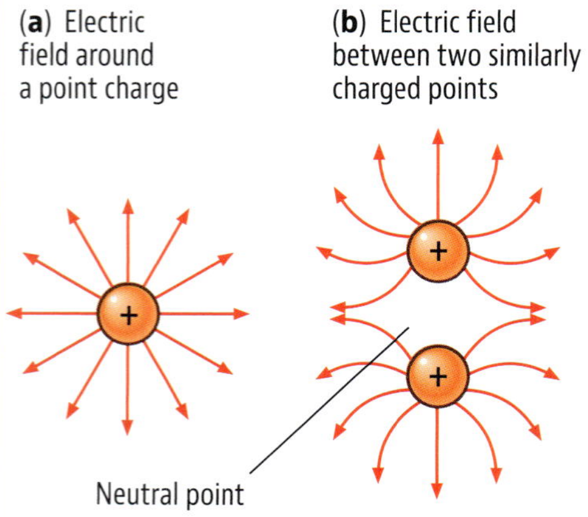

The Leyden jar, invented at the University of Leyden in 1746, was the earliest device for storing an electric charge.

The nature of the field between two charges depends on whether they are of the same sign (both positive or both negative) or of opposite sign. Similar charges repel each other, and the lines of force of the field between them are pushed apart. Unlike charges attract each other, and the lines of force are a series of curves that run from the positive charge to the negative charge.

Fields Around Objects

Any charged object has an electric field around it. The charge around a pear-shaped charged object is not even, but is largest near

SCIENCE WORDS

Capacitor: Also called a condenser, a device that can store electric charge.
Dielectric: A nonconducting material inserted between the plates of a condenser.
Electric Field: Area around a charged object in which another charge experiences a force.

SHAPES OF FIELDS

The area occupied by the electric field around a charged object depends on the shape of the object. It tends to mirror the charge distribution on the surface of the object and is strongest in the region where the charge is greatest. There is no field inside a hollow object.

(**a**) Electric field around an asymmetrical charged solid object

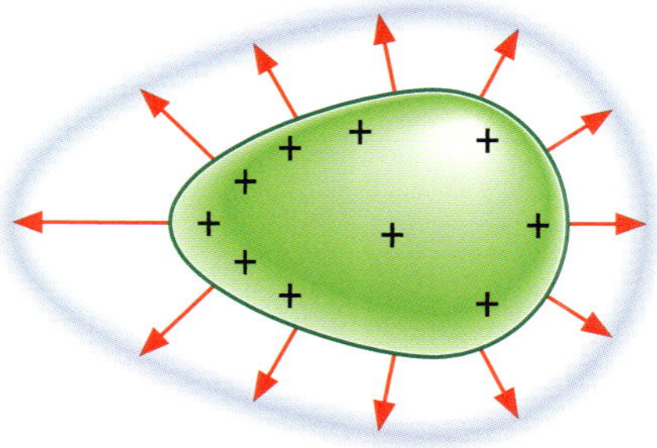

(**b**) Electric field around a hollow box

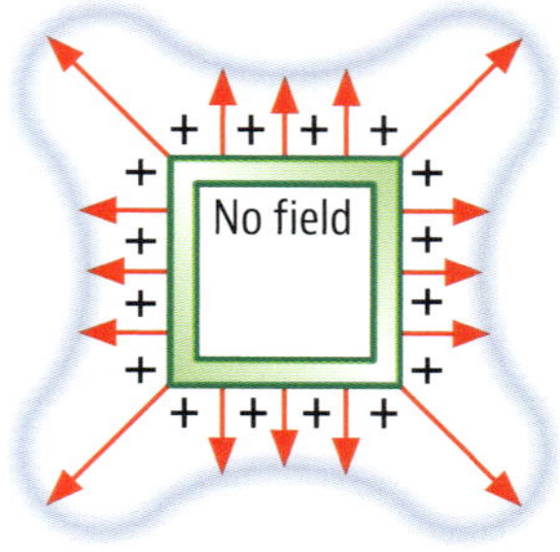

the pointed end. A hollow charged object has no field at all inside it (where there is no charge), and the external field is concentrated at any points or corners. Two charged metal plates also have a field between them. If the plates are parallel and carry opposite charges, the field is a uniform one with parallel lines of force flowing from the positive plate to the negative plate. Electric fields between charged plates are important in many electrical devices, from spectrographs to electron microscopes.

Storing Charge

Electric charge can be stored in a capacitor. The earliest type was the Leyden jar, a glass jar partly lined inside and out with metal foil. A metal knob was connected with the inner foil by a rod to which was attached loose chain, and the jar could be charged up by an electrostatic generator. In modern terms, the pieces of foil are the plates of the capacitor, and the glass jar is the insulator between them, the dielectric. A typical small capacitor, as used in radios and amplifiers, has two disks of metal separated by a ceramic insulator.

MODERN CAPACITOR

One type of modern capacitor consists of two long strips of metal foil, usually aluminum, separated by layers of waxed paper, which forms the dielectric. Connecting wires are welded to the ends of the metal strips, which act as plates. The whole arrangement is tightly wound so that it takes up the smallest possible amount of space.

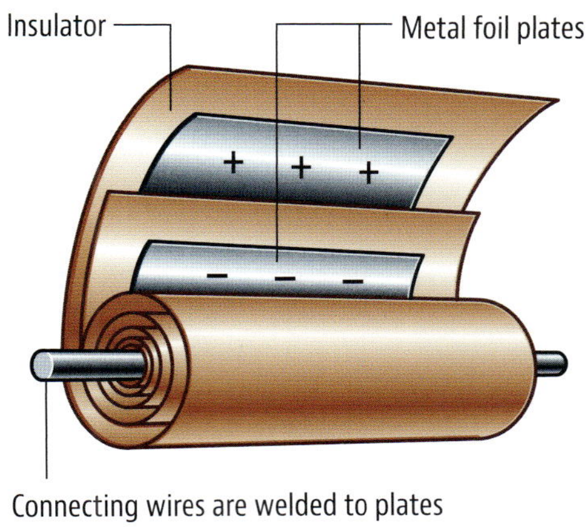

CHARGED ATOMS

An atom carrying an electric charge is called an ion. Ions behave differently from uncharged atoms. They are commonly made from elements by chemical methods, although physical processes can also be used to make ions.

An atom with a positive charge is a positive ion, or anion. Adding an extra electron to a neutral atom produces a negative ion, or cation. In other words, positive ions are atoms with too few electrons, and negative ions are atoms with extra electrons.

Hydrogen is the simplest atom, with a positively charged nucleus and a single electron. If the electron is removed, the hydrogen ion that results has a single positive charge. The outer electrons can also be removed from metals, which mostly form positive ions. Some, such as sodium, form ions with a single positive charge. Others can have more than one charge. For example, calcium forms positive ions with two positive charges, while aluminum has three.

Nonmetallic elements are better at taking on an extra electron or two to form negative ions. For example, a chlorine atom has seven outer electrons. It readily increases this number to eight to form a negatively charged chloride ion. Oxygen, with six outer electrons, takes on two more electrons to form the oxide ion, which has two negative charges.

Ionic Partners

As with all charged particles, ions of opposite charge tend to attract each other. When they do, they form a chemical compound. One of the best known compounds of this kind is table salt, used to flavor food. The chemical name is sodium chloride. In its solid state, it exists as crystals. The crystals, in turn, consist of a regular array of atoms, but the atoms carry electric charges—they are ions. In chemical terms, the molecules of sodium chloride are held together by ionic bonds between the positive (sodium) and negative (chloride) ions.

Making Ions

To a chemist, sodium chloride is a salt, one of the commonest kinds of compound. Any chemistry student will know that there are many different ways of making salts, but in the end they all involve removing one or more

CHANGING PARTNERS

A sodium atom has one outer electron, and a chlorine atom has seven **(a)**. When the two atoms react, an electron transfers from sodium to chlorine **(b)**, creating the two ions of sodium chloride.

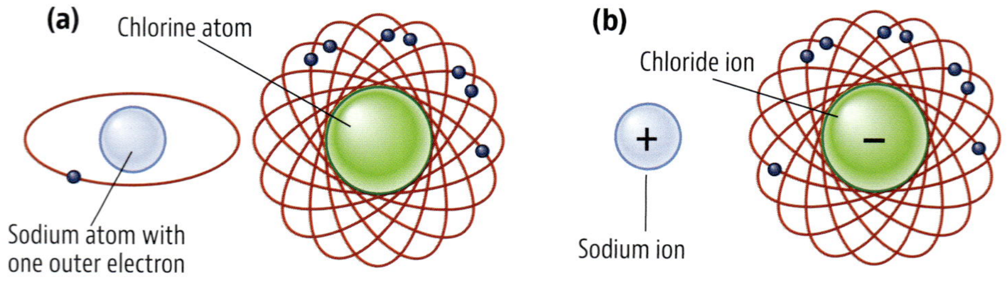

Halite is the mineral form of sodium chloride, otherwise known as common salt. It is composed of ions, and its cubic shape reflects the way the ions are arranged in its crystals.

electrons from the metallic atoms and adding electrons to the "acid" part of the salt. One simple chemical process that does this takes place when the metal is dissolved in the appropriate acid. Zinc dissolves in hydrochloric acid to produce the salt zinc chloride. Processes that seem more physical can also make ions. When a high-voltage spark (such as lightning) or ultraviolet radiation (from the Sun) passes through air, some of the oxygen and nitrogen atoms form positive ions. The electrons released immediately latch onto nearby gas molecules and turn them into negative ions. In a short time, the positive and negative ions come together and recombine, but high in the atmosphere recombination is slow. Free ions in the atmosphere create a layer of ionized gas called the ionosphere. This acts like a mirror for certain radio waves and makes long-distance communications possible. A transmitted signal can bounce off the ionosphere and be picked up by a receiver some way over the horizon.

SCIENCE WORDS

Anion: An ion with a positive charge.
Cation: An ion with a negative charge.
Ion: An electrically charged atom or group of atoms that has either lost one or more electrons (to form a positive ion, known as a cation) or gained one or more electrons (to form a negative ion, known as an anion).

ION FORMATION

When an atom loses an electron, it forms an ion with a single positive charge. When an atom gains an electron, it forms an ion with a single negative charge. Some ions have more than one positive or negative charge.

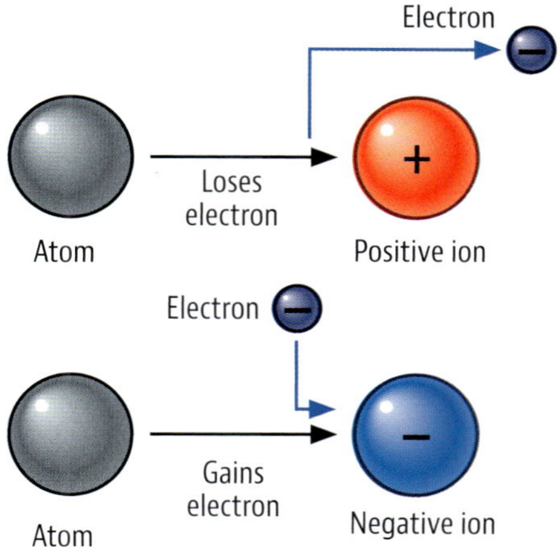

IONIC CRYSTAL

Sodium chloride crystals consist of a regular array of positive sodium ions and negative chloride ions. The sodium ions are shown in purple, the chloride ions are blue. Within the crystal the ions occupy the corners of cubes. This arrangement is evident in the shape of the crystals that are formed.

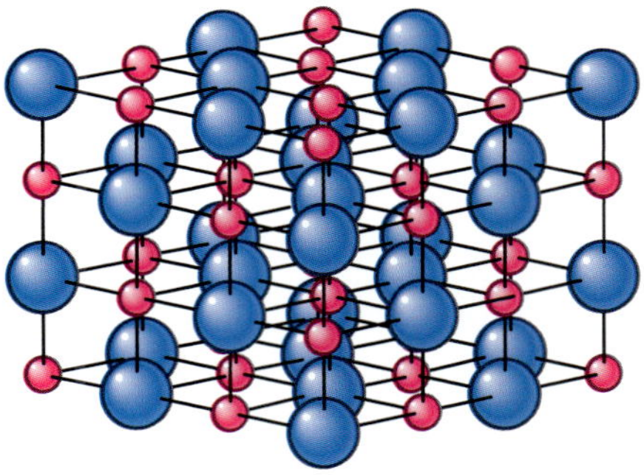

IONS AND ELECTROLYSIS

Ions can make up solid objects such as crystals, but when an ionic solid, such as salt, is dissolved in water, the ions suddenly become free to move around. The same thing happens when an ionic solid melts.

Most crystals are hard solids, because they are made up from ions that are held together by strong forces of attraction. These ionic bonds are the electrostatic forces that always exist between positive and negative charges. If you take a crystal of ordinary table salt and drop it into a glass of water it dissolves and loses all trace of its hardness and crystal structure. However, when salt is added to a jar of olive oil nothing happens—the salt just sinks to the bottom of the jar. It seems that water can split salt into its component ions, but olive oil cannot.

The reason for this has to do with the molecular structure of water. A molecule of water (H_2O) has a slight positive charge on its hydrogen atoms and an equal and opposite slight negative charge on its oxygen atom. Molecules with a slight charge separation like this are called polar molecules, and for this reason water is termed a polar solvent.

When salt dissolves in water, water molecules cluster around the ions, with their negative ends next to the positive sodium ions and their positive ends next to the negative chloride ions. This has the effect of pulling the ions away from their positions in the crystal and letting them move freely around in the salt solution. Oil does not have polar molecules. It is what is called a nonpolar solvent, and it therefore cannot dissolve salt.

To a scientist the sea is a vast solution of salt, an electrolyte that covers 73 percent of Earth.

SCIENCE WORDS

Electrolysis: The decomposition of an electrolyte by electric current, using two electrodes.
Electrolyte: A liquid (a solution or a molten substance) that conducts electricity, as in a battery or in electrolysis.
Ionic bond: A type of chemical bond formed between ions of opposite charge.

Liquids That Conduct

A substance that separates into ions when it dissolves in water is called an electrolyte. Salt is an electrolyte, but sugar is not. Although sugar dissolves, it does not form ions. Because ions can move in solution, they can conduct electricity, in much the same way as electrons (which are charged particles) can conduct electricity through a vacuum. Like salts, acids

FROM ORDER TO DISORDER

When salt dissolves in water, the sodium (+) and chloride (-) ions move away from their regular positions in the crystal and are free to move at random in the solution.

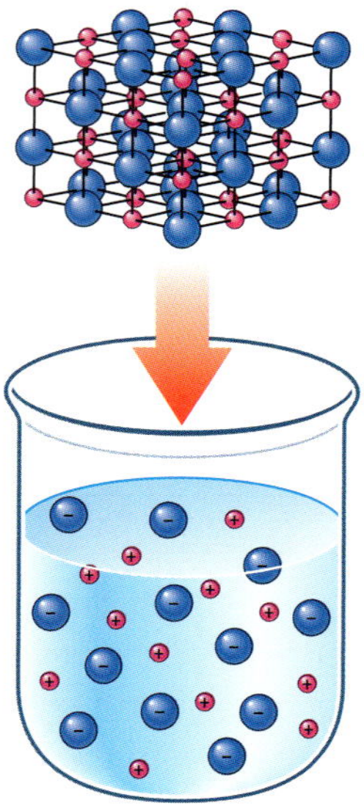

and bases (alkalis) also usually make good electrolytes. On its own, water is not a very good conductor, but if a drop of acid, which produces hydrogen ions, is added to the water it becomes an electrolyte.

When a salt is made hot enough to melt, its ions separate, and it becomes an electrolyte. Molten salts conduct electricity, and some are important electrolytes in industrial processes.

Electrolysis

In electrolysis, a pair of electrodes dip into an electrolyte and are connected to the terminals of a battery. The electrode connected to the battery's negative terminal is called the cathode, and the electrode connected to the positive terminal is the anode. Current flows from the positive terminal to the anode, through the electrolyte, and then from the cathode back to the battery's negative terminal.

TRY THIS

Potato Power

You will need some lengths of insulated wire with small alligator clips at the end, although paper clips and aluminum tape (or strips of kitchen foil) will do instead. A 9-volt battery is best because you can wind connecting wire around its terminals, or you can use tape to connect folded aluminum foil to an ordinary flashlight battery.

What to do

Cut a potato in half. Clean two copper coins with steel wool or sandpaper until they are bright and shiny. Push the coins about halfway into the cut edge of the potato, about ½ in (1 cm) apart. Connect each alligator clip to a coin, and wrap the other end of each wire around the terminals of the battery. Use a marker pen to write a plus sign (+) next to the coin connected to the positive terminal of the battery and a minus sign (-) next to the coin connected to the negative terminal of the battery. Leave the setup for about one hour, and then disconnect the battery and remove the coins. What do you notice?

There will be a green color around one of the slots in the potato—the one marked +. It was formed when copper metal from the coin reacted with chemicals in the potato to form copper salts, which are green. For this to happen an electric current had to pass through the potato to the other coin, carried by charged particles called ions.

This experiment proves that a potato can carry an electric current.

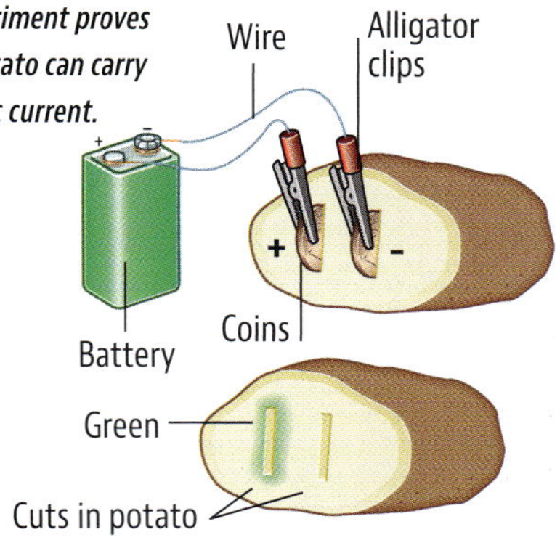

TRY THIS

Conducting Water

Liquids containing charged ions are fairly good conductors of electricity. Pure water contains few ions (charged atoms) and is therefore not a good conductor. However, by adding a few ions water conducts reasonably well.

What to do

Screw a flashlight bulb into a bulb holder and attach the wires to a 9-volt battery as shown in the illustration below. Cut two pieces of aluminum foil about 3 in x 3 in (7.5 cm x 7.5 cm), and fold each piece into three. Nearly fill a jar or beaker with water, attach alligator clips to the ends of the folded foil, and dip both pieces of foil into the water without touching each other. If water allows electric current to pass through it, the bulb will light up. However, it does not light up! Remove the pieces of foil, and dissolve 2 or 3 teaspoonfuls of salt in the water. Replace the foils and watch what happens. The bulb lights up, if only dimly. Adding salt to the water allows it to pass electricity. This is because the dissolved salt is made up of ions. Ions carry electricity through salt water solutions.

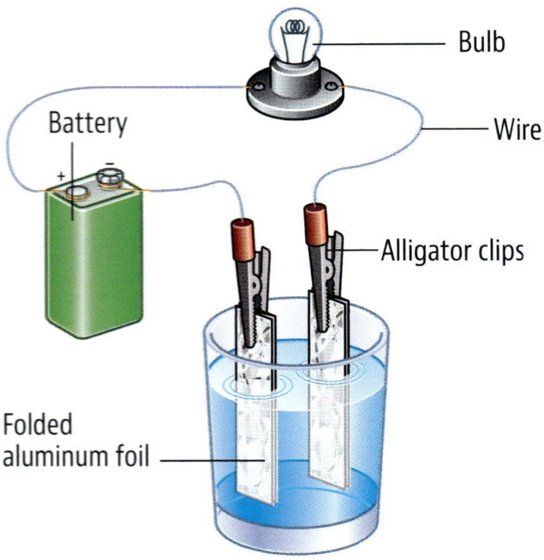

When you add salt to the water it will conduct electricity.

Within the electrolyte, negative ions are attracted to the positive charge on the anode, and positive ions are attracted to the cathode. It is this movement of ions, in both directions at once, that carries the current through the electrolyte. What happens when the ions get to the electrodes depends on the composition of the electrolyte. It may also depend on the material from which the electrodes are made.

Producing Metals

The best way to understand electrolysis is to look at some examples. An electrolyte might consist of a solution of a copper salt—say copper sulfate—in water. Copper plates are used for both electrodes, connecting them to the positive and negative terminals of a battery. In the electrolyte solution, copper sulfate splits into positive copper ions and negative sulfate ions. At the cathode (negative electrode), the copper ions acquire electrons and become atoms of copper metal. The copper is deposited onto the cathode. At the anode, copper atoms in the metal of the anode lose electrons and become copper ions—which then travel through the solution to the cathode.

If the electrodes were weighed before and after the experiment, we would find that the cathode had gained weight and the anode had lost exactly the same amount of weight. The process is used industrially to make pure copper.

Producing Gases

Another example to study is the electrolysis of water to which a little acid has been added to make it a better conductor. This time, the ions in the electrolyte are hydrogen ions (H^+) and hydroxyl ions (OH^-). The electrodes are made from the metal platinum because it is an unreactive element. At the cathode, positive hydrogen ions are discharged and

MOLTEN ELECTROLYTE

In the electrolysis of molten sodium chloride (NaCl), sodium ions (Na^+) travel to the cathode. There they acquire electrons and become sodium metal. At the same time, chlorine ions travel to the anode, where they give up electrons and become chlorine gas.

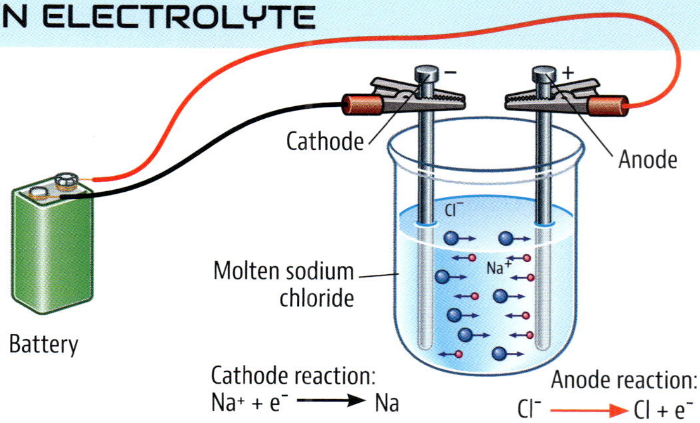

form bubbles of hydrogen gas, which rise above the liquid. At the anode, hydroxyl ions are discharged to produce water molecules and bubbles of oxygen, which also rise above the liquid. There will be exactly twice as much hydrogen as oxygen in accordance with water's chemical formula, H_2O.

Uses of Electrolysis

Jewelry, knives and forks, automobile parts, and faucets and other plumbing fittings are among the hundreds of articles that are often electroplated. Jewelry made from an inexpensive metal such as brass can be gold-plated to make it look more beautiful and to prevent it from tarnishing. Cutlery made from an alloy of nickel and copper is silver-plated for the same reason. Such cutlery is often marked "EPNS," which stands for "electroplated nickel silver". Fenders and other automobile parts made from steel are chromium-plated to make them look good and to prevent the steel from rusting. Often copper and then nickel are plated onto the steel first, before the final thin layer of chromium. Faucets and other bathroom fittings, usually made from brass, are also often chromium-plated to improve their appearance and corrosion resistance.

GASES FROM WATER

The electrolysis of acidified water produces the gases hydrogen and oxygen. At the cathode, hydrogen ions gain electrons and form hydrogen gas. At the anode, hydroxyl ions lose electrons, forming water molecules and oxygen gas. The proportion of hydrogen atoms to oxygen atoms is 2 to 1, as expected from the formula H_2O.

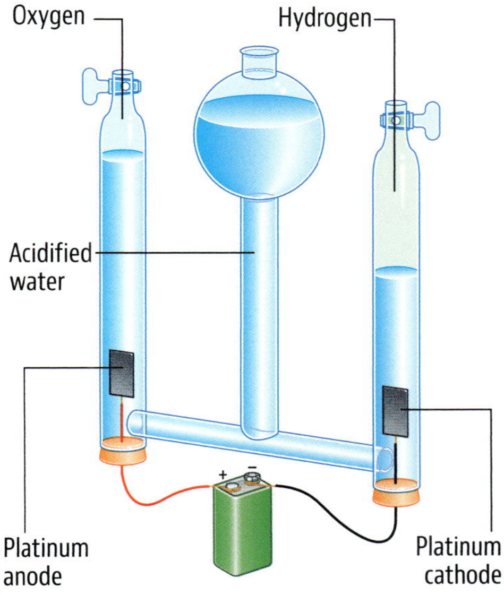

MOVING CHARGES

An electric current consists of electrically charged particles in motion. Such currents played a vital role in the development of modern civilization, creating heat, exerting magnetic forces, and carrying messages.

All matter is made up of electrically charged particles. Every atom has a central core, or nucleus, that contains most of its mass and is positively charged. And moving around the nucleus are much lighter, negatively charged particles called electrons.

The forces between electrically charged particles are very strong. These powerful forces hold the atom together. But when equal amounts of positive and negative charge are close to each other, the effects of the charges cancel each other out, and a short distance away it is as if no charge at all were present. In normal atoms, the positive charge on the nucleus is exactly balanced by the negative charge of the electrons. Electrical effects become noticeable when electrons are removed from atoms, leaving ions (an atom or group of atoms with one or more electrons added or removed) that have an unbalanced positive charge.

Electric Charges

When you comb your hair with a plastic comb, you sometimes see strands of hair attracted to the comb. This happens when electrons have been transferred from the comb to the hair, leaving positively charged atoms behind. The negative charges on the hair and the positive charges on the comb attract each other. This is an example of what is known as static electricity.

Electric charge also shows its presence when it moves along a conductor as an

ELECTRONS ON THE MOVE

Metals are good conductors. Electrons can easily become detached from their atoms and form a "sea" of electrons throughout metal wire (below, left). When a voltage is applied along the wire (below center) the electrons drift along together in the same direction as an electric current. If the same voltage is applied to a wire of different metal, a different current flows. If the current is smaller—that is, if fewer electrons move—the metal is described as having a higher resistance than the first one (below right). Heat is generated whenever a current flows through a conductor. The heat generated depends on the resistance of the conductor and on the amount of current flowing through it—the higher the resistance, the greater is the amount of heat.

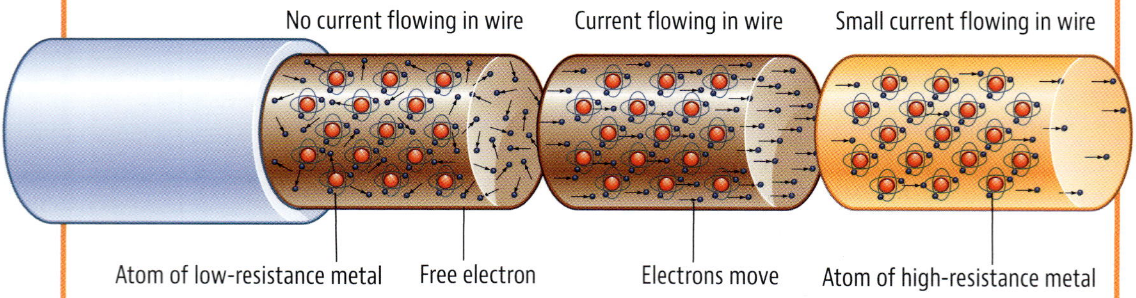

A pylon supporting a high-voltage cross-country power line. The current generated by national electricity supply networks is used at a much lower voltage in people's homes.

electric current. When a light or television set is switched on, electrons flow along the wires connecting the appliance to the electricity outlet in the wall and in the wires and cables connecting the outlet to the power plant. When a pocket calculator is switched on, very weak electric currents flow in the metal connections and the microchips inside the calculator.

Everyday Uses

Electric currents are important because they have so many different uses in our daily lives. They heat up the wires through which they flow, so they can be used in electric heaters, electric irons, and electric lamps. The currents also have magnetic effects. For example, rapidly varying electric current in a loudspeaker pulls on magnets in the loudspeaker's moving parts, which oscillate rapidly to produce sounds. Electric currents can also be used for signaling. Currents carry signals that represent sound to and from your telephone, they carry data in your computer, and they carry the information from which pictures are built up in your television set and other electrical devices.

TRY THIS

Can It Conduct?

Substances that allow electric current to flow through them are called conductors. Other substances that do not allow electric current to flow through them are called nonconductors or insulators. This experiment tests various substances to find out if they are conductors or insulators.

What to do

Screw a flashlight bulb into a bulb holder then attach lengths of wire to the bulb holder and a battery as shown in the illustration below. Test a selection of different materials one at a time by clipping alligator clips to them. The article completes a circuit. If it allows current to flow through the circuit, the bulb will light up. This means the article is a conductor. If the bulb does not light up, the article is an insulator. Put each article into one of two groups, one group for all the conductors and the other for all the insulators. The conductors will all be metals, such as coins, a paper clip, a metal spoon, a safety pin, and a steel nail. All metals conduct electricity.

The insulators are a wide variety of materials, but are all nonmetals. They include items such as paper, an eraser, a rubber band, a plastic spoon, a wooden ruler, a piece of string, and a pencil. Paper, rubber, plastic, string, and wood are all insulators—plastic is commonly used to insulate electrical wiring.

Test each item to see if the bulb lights up.

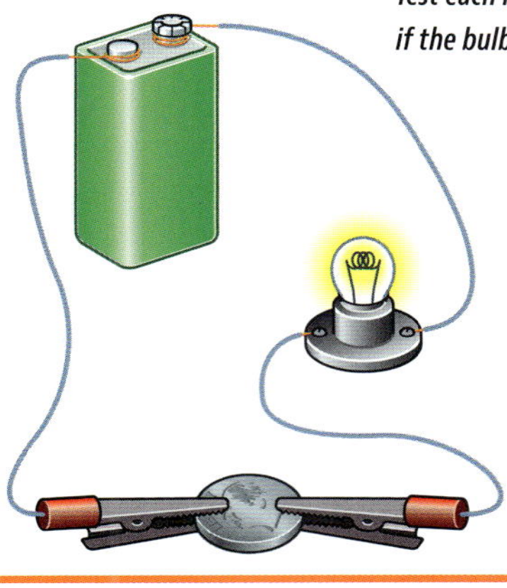

POTENTIAL DIFFERENCE

Electric charges need a push to make them move. This push is called potential difference and it is measured in volts. Power plants and batteries both create potential difference, on very different scales, to get electric currents flowing.

The physical forces exerted in combing your hair can move small amounts of charge. Pulling off a shirt made from artificial fibers can have the same effect: you feel a tiny shock, and sometimes if it is dark you can see sparks. When you walk on certain types of carpet, charge builds up on your body. You notice this when you touch a metal object, such as a faucet, and feel a small shock as the charge flows into it.

Charge flows easily through metals. These are called electrical conductors. It can hardly flow at all through other materials, such as rubber and most plastics. These are called electrical insulators. For example, the cord to a desk lamp is made from copper wires (along which the current flows) that is coated with plastic (through which the current cannot flow).

If you comb your hair with a metal comb you will not get a shock or see any sparks. Electrons that are dislodged from your hair immediately flow away through the metal and cannot build up into a sizeable quantity of charge, as they can on a plastic comb.

Making Currents Flow

An electric battery is a way of making currents flow. When the two terminals are connected through conducting wires to a device such as

In an electric train, current flows from high-voltage overhead lines and through the pickup gear on top of the train. Electricity travels through electric motors, turning shafts that drive the wheels.

a flashlight bulb, an electric current is driven through the device. In chemical reactions that take place inside the battery, electrons are separated from their atoms. The electrons are forced through the wires and then through the device. When the device is disconnected the electrons cannot move, stopping the chemical reaction.

Power plants use more powerful devices for making electric currents flow. Steam, generated using the heat of burning coal, oil, or gas, or from nuclear energy, is used to drive huge turbines. Electric generators coupled to the turbines produce high-voltage current that is distributed along cables that run across country.

In a wire in which electric current is flowing, there are trillions of electrons in motion. But the electric charge is zero overall. The negative charge of the electrons is canceled out by the positive charge on the atoms that have lost their electrons.

SCIENCE WORDS

ampere (A): The SI unit of electric current. A current of 1 ampere (often abbreviated to "amp") is equal to a flow of 1 coulomb per second.
Potential difference: Also called voltage, the difference in electric potential between two points. It is measured in volts. The higher the potential difference, the greater the force tending to move charges between the points.
volt (V): The SI unit of potential difference.

If the circuit is broken, electrons immediately stop flowing. If they momentarily began to accumulate at the break or any other point in the circuit, their electric charge would repel others following behind, and instantly disperse the accumulation of charge. This is why electric current cannot flow unless there is an unbroken loop, a circuit, that it can follow.

The "push" that drives electrons around a circuit is called potential difference, or p.d. Another name is "voltage" because p.d. is measured in volts. Ordinary flashlight batteries provide a voltage of about 1.5 volts; automobile batteries, 12 volts; and a domestic outlet for a lamp, about 110 to 130 volts (in North America and some other places), or between 220 to 240 volts elsewhere. Still higher voltages are used to send electricity across country from the generating plants to factories, offices, and homes.

Resistance

Most materials are neither perfect conductors nor perfect insulators. They resist the flow of current to a greater or lesser extent. A piece of wire included in a circuit to control the flow of current by its resistance is called a resistor. Resistance is measured in units called ohms.

VOLTAGE, RESISTANCE, AND CURRENT

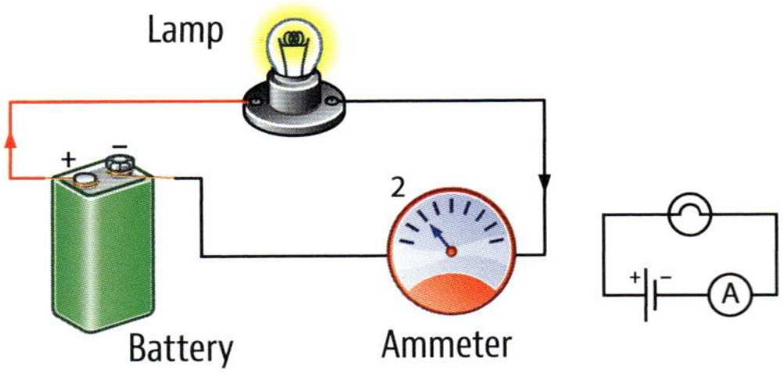

A battery and ammeter (for measuring current), both with zero resistance, are connected in series with a bulb so that the same current flows through all three. The ammeter shows that the current is 2 amps.

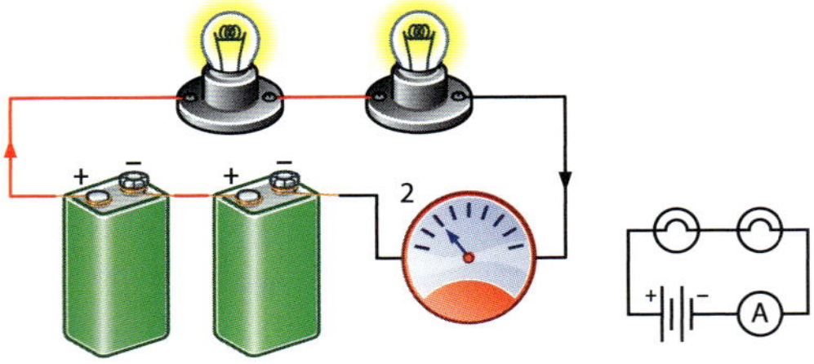

When two bulbs are connected in series their combined resistance is twice that of a single lamp. The current is halved to 2 amps.

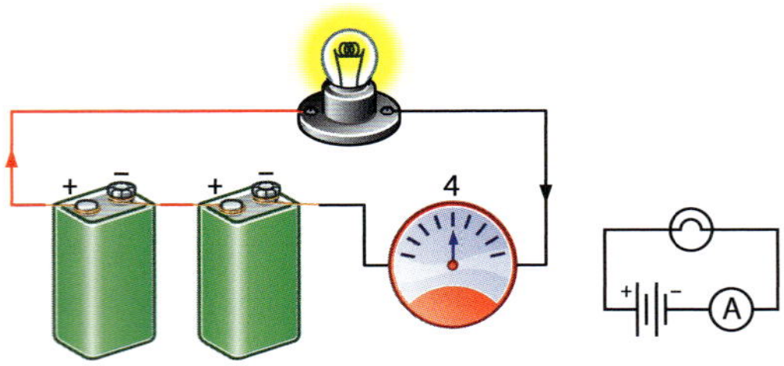

Two batteries in series give twice the potential difference across the bulb. With one bulb, the resistance in the circuit is unchanged, the current is now 4 amps.

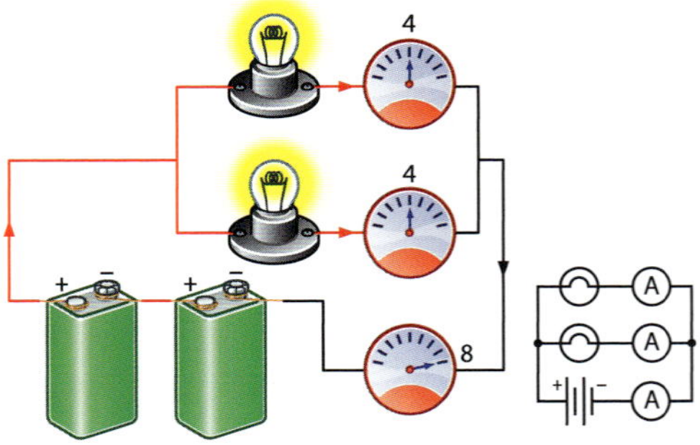

Two bulbs are connected in parallel, but each has the potential difference of two batteries across it, so a current of 4 amps again flows through each. The combined current is 8 amps.

The greater the resistance of a component in a circuit, the more p.d. is needed to make a given current flow in it. In a small pocket flashlight the bulb may have a resistance of about 3 ohms and two 1.5-volt batteries. (The total p.d. that can be delivered by a voltage source, such as a battery or generator, is called its electromotive force, or e.m.f.) The batteries are connected end-to-end so that their e.m.f.s add together to give a total voltage of about 3 volts. Current is measured in amperes—"amps" for short. The current in amps that will flow through the bulb is given by the e.m.f. in volts divided by the resistance in ohms, in this case 1 amp. (It is actually slightly less than that because the batteries and other components in the circuit also have some resistance.)

Measuring Current

The amount of current that will flow in a circuit depends not only on what components are connected into the circuit, but on the ways in which they are connected. In the diagram above, first one and then two identical batteries are used to light a bulb. Then the two batteries are used to light two bulbs. Current-measuring devices

SCIENCE WORDS

Parallel: Two electrical components in a circuit are connected in parallel if the current divides to pass through them separately. See also Series.
Resistance: A measure of how a material or a component resists the passage of electric current through it. The higher the resistance, the less current will pass when a given potential difference is applied across it.
Resistor: An electrical component with a known resistance, used to regulate current and voltage in a circuit.
Series: Two electrical components in a circuit are connected in series if the same current passes through both of them in turn. See also Parallel.

called ammeters are included in the circuits. The resistances of the batteries and the ammeters are so small that they can be ignored.

Series and Parallel

The bulbs are first connected in series—that is, so that the same current passes through both of them. Their resistances add, so that their combined resistance is twice the resistance of one of them. Less current flows than when there is only one bulb in the circuit.

When the bulbs are connected in parallel, the current is split as it passes through them. There is the same voltage across each bulb, and it produces the same current in each bulb separately. The currents join as they flow out of the bulbs, so the combined current in the main part of the electrical circuit is twice the current there would be if only one bulb were in the circuit. The combined resistance of the two bulbs in parallel is effectively equivalent to half the resistance of each bulb singly.

TRY THIS

Series or Parallel?

There are two main ways of connecting components in an electric circuit. They can be chained, like beads on a necklace, so that the same current flows through them all, one after the other. Components connected like this are in series. Alternatively, one wire connects all the contacts on one side of each component and attaches them to one terminal of the battery. Another wire connects the contacts on the other sides of the components to the other battery terminal. Components connected like this are said to be in parallel.

What to do

The first diagram below shows the series circuit. The diagram to the right is the parallel circuit. Connect the series bulbs. Note how they look. Then connect the parallel holders. Do the bulbs look any different? In the series circuit, the current had to pass through each bulb in turn. As a result, the bulbs were not very bright. In the parallel circuit, the full current from the battery went through each bulb. All were as bright as if only one was being lit. However, there is a price to pay—the battery will run down three times faster than it would with the series arrangement.

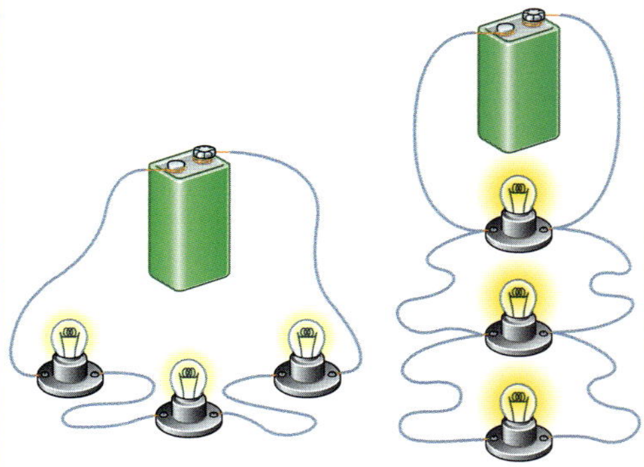

Bulbs wired in series have to share the voltage between them—the bulbs are not bright.

Each of the bulbs wired in parallel receives the battery's full voltage—the bulbs are bright.

CURRENT AND STORED CHARGE

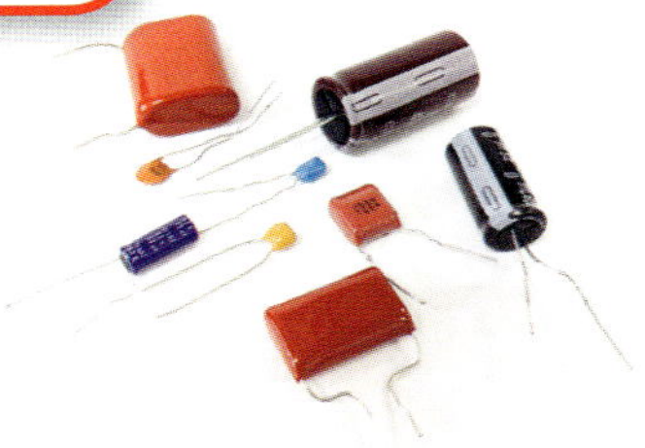

A selection of different capacitors.

To make use of electric current, it needs to be measured accurately. A device for measuring current is called an ammeter. Nearly all measuring instruments make use of the current's magnetic properties. Often it is necessary to stop the flow of current and store charge in one place, in a device called a capacitor.

Many current-measuring devices rely on the fact that an electric current sets up a magnetic field around itself. This field will move a nearby compass needle and push or pull a nearby wire carrying another current. If a wire is looped into a coil, it behaves like a magnet, with one end of the coil acting as the magnet's north pole and the other as the south pole. If such a coil is hung from a thread and a current is passed through it, the coil will swing so that its poles are pointing north–south, like the needle of a magnetic compass. The stronger the current flowing in the coil, the stronger the twisting force.

The Moving-Coil Ammeter

In the usual type of ammeter, the current flows through a coil that is wound around a core of soft iron (that is, nearly pure iron, rather than steel). The core "magnifies" the coil's field. The coil is suspended between the two poles of a strong permanent horseshoe-shaped magnet. When current flows, the coil twists to line up with the permanent magnet's field, but the spring by which it is suspended resists this motion. The coil turns through a greater or smaller angle, depending on the strength of the current, which is indicated by a needle attached to the moving coil and a scale.

The Mirror Galvanometer

A very sensitive current-measuring device that does not use a moving needle is called a mirror galvanometer. Again it uses a coil that is suspended in the magnetic field of a permanent magnet. It is also enclosed within a draft-excluding container with a window. The coil carries a mirror, which reflects a beam of light shone in through the window. As a current flows through the coil, the coil swings backward and forward through an angle that is larger for larger currents. Several yards away the beam leaving the window forms a spot of light that moves first one way and

CHARGING A CAPACITOR

A battery transfers electrons from one plate of a capacitor to the other. Their mutual attraction holds the electrons there when the battery is removed. The plates discharge, causing an electric spark, when the wires are brought together.

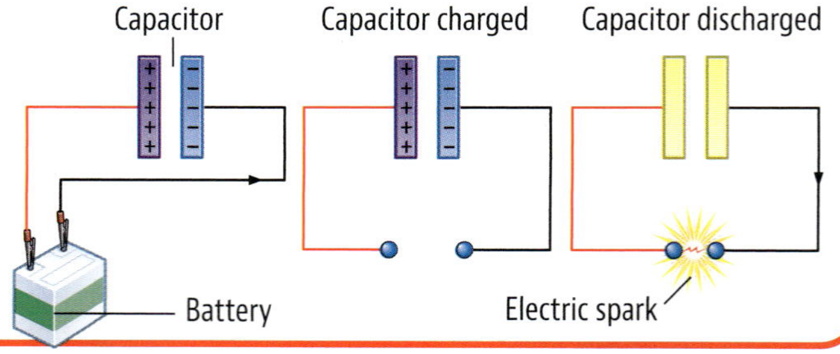

SCIENCE WORDS

Ammeter: An instrument for measuring electric current.
Capacitance: Also called capacity, the ratio of the stored electric charge on an electrical device or other object to the voltage applied to it.
Magnetic Field: The pattern of magnetic influence around an object.

then the other through a long distance, making accurate measurements possible.

Storing Charge

A capacitor stores electric charge. A highly simplified type consists of two parallel metal plates. If they are connected to the two terminals of a battery, electrons pushed away from the negative terminal accumulate on one plate. Electrons are drawn away from the other plate by the attraction of the positive battery terminal. As electrons build up on the first plate, they repel following electrons increasingly strongly, and the current falls. Electrons flowing from the second plate toward the battery are held back by the attraction of the net positive charge remaining on that plate. The current falls to zero. If the battery is removed and the two ends of the wires from the plates remain unconnected, the electrons stay on one plate. If the two wire ends are brought close together, the electrons can flow across the small air-gap to the other plate, attracted by the positive charges. There is now no net charge on either plate—the capacitor has been discharged.

Capacitors are key components in circuits. One type consists of two long strips of metal foil, separated by a material called a dielectric and rolled up. The dielectric increases the amount of charge that can be stored.

MOVING-COIL AMMETER

A current through the coil temporarily turns it into a magnet. Affected by the field of the permanent magnet, it rotates, moving the needle over a scale.

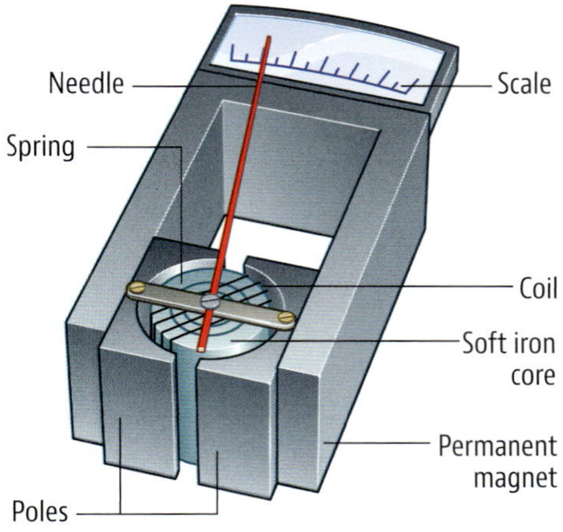

MIRROR GALVANOMETER

The moving coil moves almost without friction in a draft-excluding chamber. It carries a mirror that reflects a light beam to a large distance, amplifying the smallest movement.

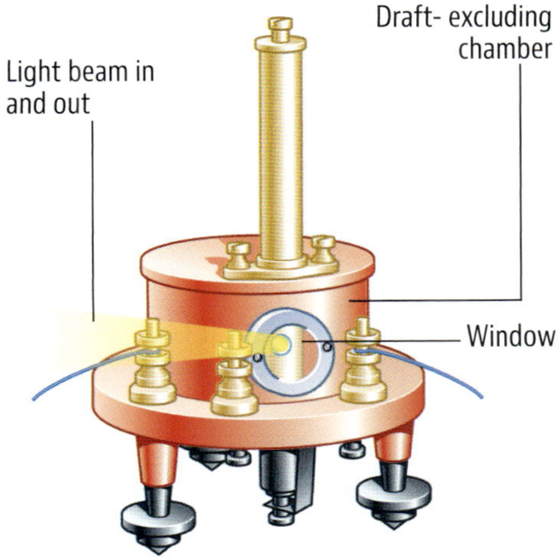

RESISTANCE AND POWER

The power needed to drive an electrical device, and the power that it can deliver, both depend crucially on its resistance, as well as on the voltage—that is, on the potential difference (p.d.)—applied across it. Working out how much resistance to use, and where, is an important part of designing electrical circuits.

German physicist Georg Ohm (1789–1854) made an important advance in the study of electricity in the early 19th century. He found that in many materials, especially metals, the current that flows through a given piece of the material—a piece of wire, for example—is proportional to the voltage across the material. That is, if a p.d. of 10 volts is applied, twice as much current will flow as when only 5 volts is applied, provided the temperature is kept constant.

No material follows this law exactly, but those that do obey it approximately are sufficiently numerous to be very important and are called ohmic. The law can be written as:

$$I = V/R$$

in which V is the p.d., I is the current, and R is simply the resistance of that particular piece of material. This equation can also be written in the equivalent forms

$$V = IR$$

and

$$R = V/I$$

The flow of current through a resistive material generates heat. To keep the temperature constant the heat must be continually removed. As the temperature of a material is raised, its resistance generally increases, although again this is not true of all materials.

The resistance of a particular piece of material depends not only on what it is made from, but also on its shape. For example, the resistance of a metal wire is much greater than the resistance of the same piece of metal if it is melted down and formed into a shorter, fatter cylinder. For a given material, the resistance increases as length increases, and decreases as the cross-sectional area increases.

Resistors used in electrical and electronic circuits are made from metal wire or from carbon in a casing. They have their actual resistance marked on the casing in a code consisting of colored bands.

In Series and in Parallel

Resistors can be arranged in various different ways in a circuit to produce different effective combined resistances (see the diagram on

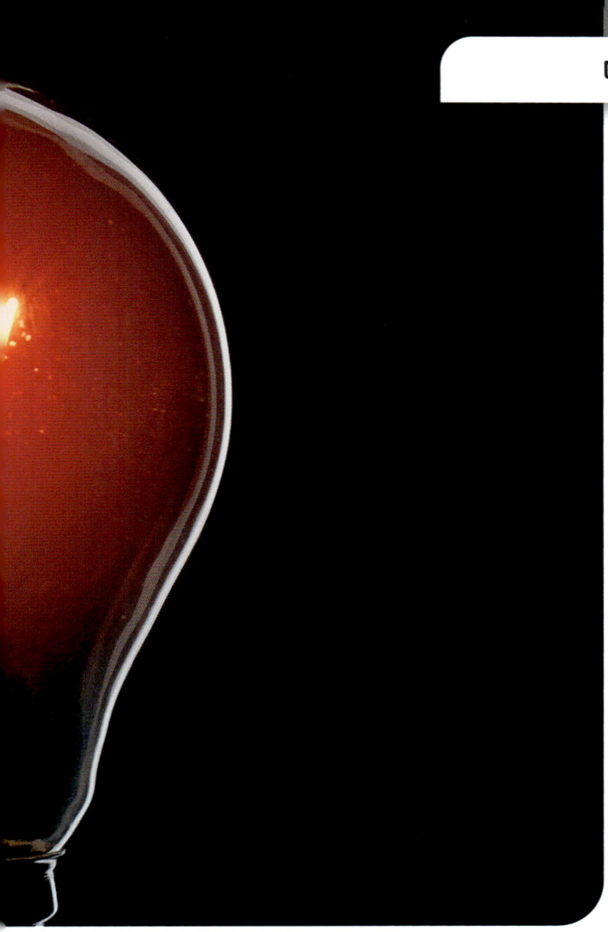

Old-fashioned incandescent light bulbs contain a wire filament that is heated until it glows. Electric current is supplied to the filament by wires.

page 36). Where two or more resistors are arranged in series so that the same current flows through all of them, their resistances add up. It needs a higher p.d. to drive a given current through them.

Alternatively, resistors can be arranged in parallel so that they have the same p.d. across them, but different currents flow through them. In this case the separate currents through the different resistors combine when they emerge. The parallel resistors effectively have a smaller resistance than any one of them does by itself.

If, for example, there are three resistors, R_1, R_2, and R_3, then the current through R_1 is V/R_1, and similarly for the other two. The

TRY THIS

Briefly Bright

Incandescent light bulbs contain a piece of very thin wire called a filament. When electricity flows through the wire, the wire becomes very hot and gives off light. A gas that does not support burning (such as argon or nitrogen) fills the bulb and prevents the filament from burning away. This project explains how to make an electric bulb. However, because this "bulb" will contain air, it will light up for only a very brief time.

Warning:
do not to touch the jar, nails, or alligator clips without wearing protective gloves. After you connect the wires the apparatus will become hot.

What to do

Push two 3-in (7.5 cm) nails through a small disk of Styrofoam, with the heads just sticking up. Tease a strand of fine wire from some steel wool and wind its ends around the pointed ends of the nails. Place the Styrofoam disk inside a glass jar as shown below. Connect alligator clips to the heads of the nails. Connect the other end of one wire to a battery. Watch carefully as you touch the end of the other wire onto the other battery terminal.

The thin wire between the nails acts like a filament in an electric lamp. When you passed electric current through it, it rapidly became hot but soon burned away. For a brief moment the jar was lit up by electric light.

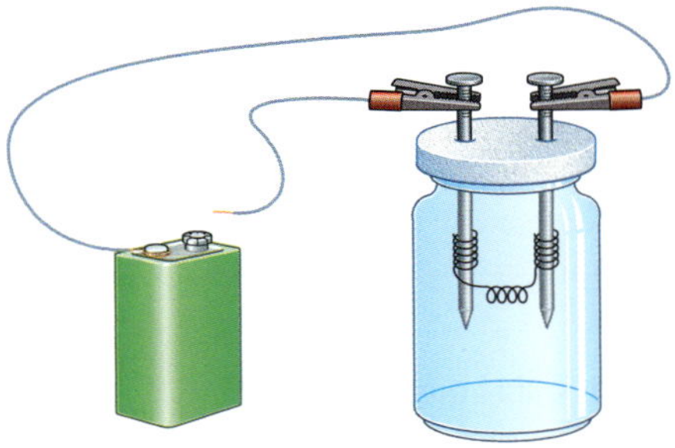

combined current is

$$V/R_1 + V/R_2 + V/R_3$$

or

$$V(1/R_1 + 1/R_2 + 1/R_3)$$

This is the current that would be produced by a single resistance R if its value is given by the equation

$$1/R = 1/R_1 + 1/R_2 + 1/R_3$$

So R is the combined resistance of the three resistors when connected in parallel.

Resistance and Heating

When a current flows through a material, it generates heat. It is put to work in electric ovens, irons, toasters, kettles, and water-heaters, in which current flowing through a wire makes that wire hot. Some electric lamps also work on this principle. The tungsten filaments (wires) in old-fashioned incandescent light bulbs are designed to get so hot that they glow and give off light.

Power and Energy

Every electrical device requires energy to make it work. A great deal is needed by a water heater, very little by a portable radio.

COMBINING RESISTANCES

When resistors are connected in series, their resistances add up. When they are connected in parallel, their combined resistance is less than any one of them individually, in accordance with the equation. If all the bulbs have the same resistance (bottom), the current will divide as shown.

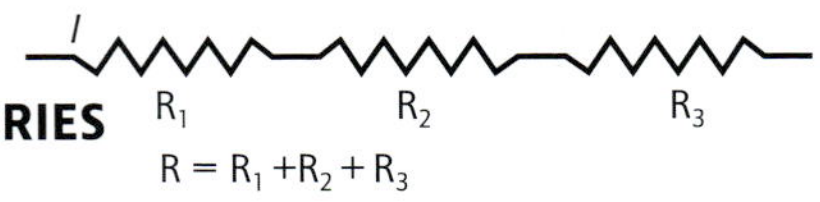

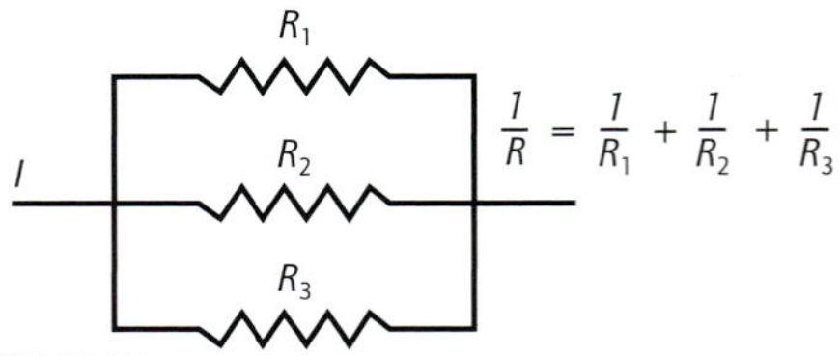

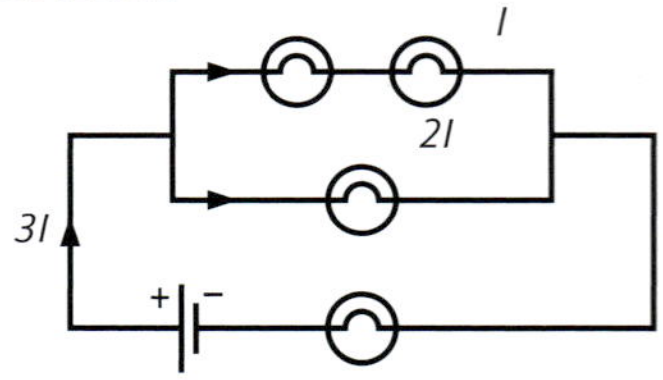

SCIENCE WORDS

ohm (Ω): The SI unit of resistance.
Resistance: A measure of how a material or a component resists the passage of electric current through it. The higher the resistance, the less current will pass.
Resistor: An electrical component with a known resistance, used to regulate current and voltage in a circuit.
Superconductivity: The property of conducting electricity with no resistance at all. Some metals do this when cooled close to absolute zero (–459.67°F / –273.15°C). New complex substances have been developed that superconduct at higher temperatures (though not yet as high as 32°F / 0°C).

The rate at which a device uses energy is called its power consumption (power is the rate of expenditure of energy). In the electrical industry, power is measured in units of watts (symbol W) or kilowatts (symbol kW). A kilowatt is equal to 1,000 watts. An energy-saving lamp consumes about 10 W, a toaster typically uses 1 kW, and a TV set about 1/6 kW.

The amount of energy a device uses in a given time is found by multiplying its power consumption by the time. The unit used is the kilowatt-hour (also simply called a "unit"). It is the energy used by a 1-kW device running for 1 hour.

Electricity consumers pay the generating companies for the amount of energy they use.

A meter on the premises measures the energy used by measuring the amount of current going to the premises. The power used is the current multiplied by the voltage at which it is supplied. It is continually measured and added up to give the total energy used.

Changing Resistance

The heat generated when an electric current flows is caused by the current overcoming the resistance of the material through which it is flowing. The material contains imperfections and irregularities in its crystal structure, and they disturb the motion of the electrons. The more perfect a crystal—meaning the fewer the imperfections in its array of atoms—the less resistance it has because electrons can flow through it freely. The heating caused by a current flowing through a material with electrical resistance is called resistive heating.

The main use of restrictive heating is in electric heating elements used in things such as electric kettles and showers. The element gets hot and heats the water as current flows through it.

Vanishing Resistance

Some materials lose all their electrical resistance, becoming superconductive, when they are cooled sufficiently. This was discovered in 1911 by a Dutch physicist, Heike Kamerlingh Onnes (1853–1926), who found that the metal mercury becomes superconductive when its temperature is less than 4 degrees above absolute zero. Other materials show the same tendency at various low temperatures. From 1986, new materials were developed that showed superconductivity at much higher temperatures, around 100 degrees K (–280°F / –173°C) above absolute zero. If superconductivity at room temperature can be achieved, very cheap power transmission will become possible because little energy will be lost from heating the transmission lines. Even faster computers could be built, and many other sophisticated new devices could be developed and made available to industry and private citizens.

TRY THIS

Electric Heater

The previous project made a short-lived electric lamp. The filament burned away because it became extremely hot. In this experiment, a thicker piece of metal is used to demonstrate the principle of the electric heater.

What to do

Cut a strip of aluminum foil about 6 x 1 inch (15 x 2.5 cm) and fold it in half twice lengthwise. The folded foil is the element of the heater. Curve the element into a shallow U shape and, using one hand, hold it against the ends of the battery (as shown below) and count to 10 slowly. If you touch the element with your other hand you will feel it getting hot. Remove the element from the battery immediately before it becomes any hotter —it will get very hot indeed.

Like all metals, aluminum is a good conductor of electricity. When it is thin, like the foil in this project, the passage of electric current through it heats it up. You could not make an electric room heater with an aluminum element since it would get too hot and burn away. Instead, room heaters have elements made of metal alloys that only become red hot and do not burn up.

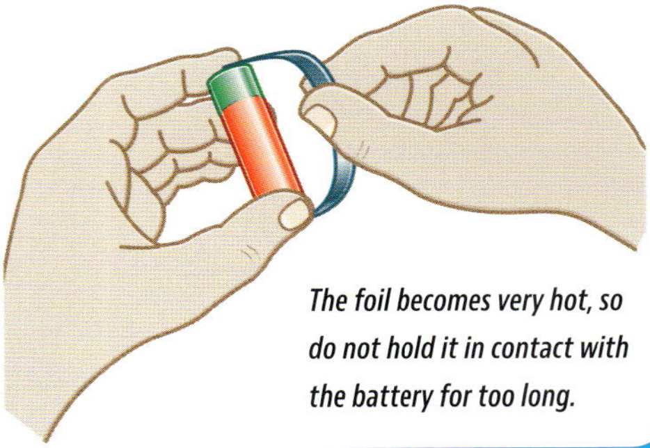

The foil becomes very hot, so do not hold it in contact with the battery for too long.

DIRECT AND ALTERNATING CURRENT

When the first power plants began generating electricity, they supplied it to homes and factories as one-way direct current. Now, all outlet current is alternating current (AC), reversing its direction constantly. This is necessary to make power controllable, so that it can be delivered at the voltages that are needed.

SCIENCE WORDS

Alternating current (AC): Electric current that flows first in one direction, then in the other, alternating many times each second. AC is used for domestic electricity supply and other electrical applications.

Direct current (DC): Electric current that flows in one direction all the time, although it may vary in strength.

Transformer: A device that increases or decreases the voltage of alternating current.

The electricity that is supplied by utility companies to homes, offices, and industrial plant is not only at a much higher voltage than the electricity supplied by batteries in a portable radio or flashlight. It also differs in another crucial way. The current from a battery is direct current (DC). This flows in one direction. The current from the regional power grid repeatedly changes its direction as alternating current (AC).

Alternating current generally has advantages over direct current for large-scale uses. AC current has strong magnetic effects, which are vital to a host of electrical devices. Related to this is the fact that AC voltage can be altered easily, but DC voltage cannot. Different voltages are needed for different purposes. AC voltages are also easily produced by turbine generators.

Changing Voltages

Every electric current generates a magnetic field. If wire is looped into a coil, the whole coil behaves like a bar magnet. An AC current generates an alternating magnetic field. An electric current can also be generated by a magnetic field, but only by a changing one. Moving a bar magnet near a wire causes a

DC AND AC

A battery delivers a one-way voltage that is steady. This produces a constant one-way current. It is called direct current (DC). The graph of voltage against time is a horizontal straight line. Turbines in power plants deliver a voltage that constantly reverses in direction, creating an alternating current (AC). The voltage–time graph then has a wavelike shape.

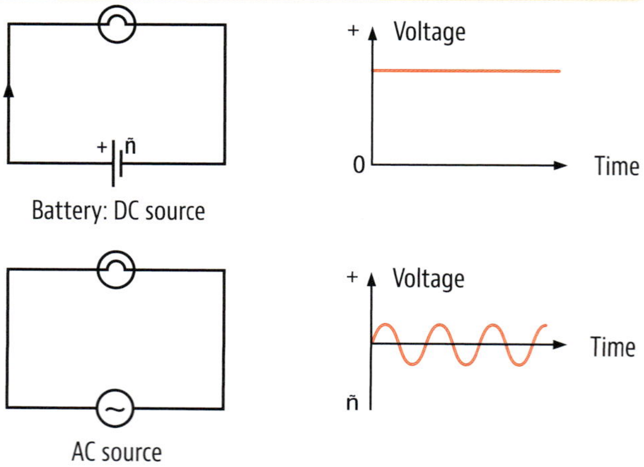

Large transformers at an electricity substation step down the voltage of the supply from the large voltages of the long-distance supply to the lower voltages used in factories or homes.

voltage to develop along the wire while the field is changing. This changing field induces a voltage. If the wire is looped into a coil of many turns, the voltage is generated across each turn of the coil. The cumulative voltage across the whole coil is increased.

This makes it possible to alter AC voltages easily. AC current flows through a coil in one circuit, called the primary circuit. The coil is wound around an iron core, which also passes through the coil of the secondary circuit. The magnetic field generated by the primary coil creates a bigger magnetic field in the iron. This field fills the interior of the iron and passes through the secondary coil. Because the field is varying in strength and direction, it induces an AC voltage in each loop of the secondary coil. The more loops there are in the secondary coil, the bigger the voltage that develops.

If there are fewer turns in the secondary coil than in the primary coil, a smaller voltage is developed in the secondary than in the primary. This device is called a transformer (see illustration above right). Transformers are used to step up the voltage of current from a power station to transmit over long-distance high-voltage lines, which run cross-country suspended from tall steel pylons. Other transformers, housed in unmanned installations called substations, are placed near industrial plants and near residential areas to step down voltages.

HOW DOES A TRANSFORMER WORK?

STEP-UP

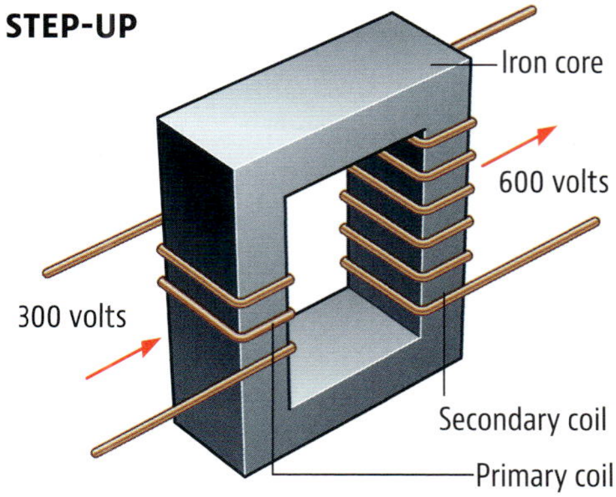

The alternating magnetic field of the primary coil is channeled through the secondary coil by the iron core. The loops in the secondary coil develop a higher voltage.

STEP-DOWN

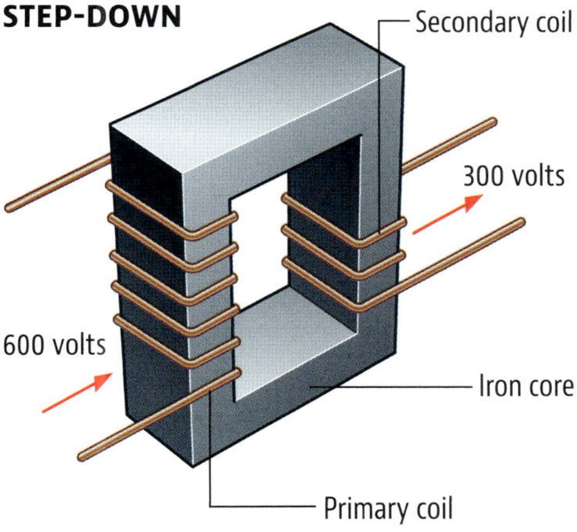

When the secondary coil has fewer loops than the primary coil, it develops a lower voltage, making the setup into a step-down transformer.

VACUUM TUBES AND THEIR USES

Before transistors, vacuum tubes like these were vital components in electronic devices. The earliest computers used many thousands of these tubes.

By the beginning of the 20th century electricity had long been used in industry. Devices called vacuum tubes offered new ways of controlling the flow of electrons. They ushered in the electronics age, without which modern life could not exist.

In the early 20th century, vacuum tubes such as the Crookes tube (see pages 6–7) evolved from being just scientific instruments into practical tools performing many different tasks. The simplest vacuum tube is called a diode and contains two electrodes—a cathode and an anode. The cathode is heated to give off electrons. When the anode is at a positive voltage in relation to the cathode, it attracts the electrons. When the anode is at a negative voltage in relation to the cathode, electrons cannot flow to it. Being cold, the anode does not give off electrons, so electrons cannot flow from the anode to the cathode.

TWO KINDS OF VACUUM TUBE

A diode contains a heated cathode, or electron-emitter, and an anode, or electron-receiver. In a triode, a third electrode called a grid controls the flow of electrons. The schematic symbols show how the diode and triode are represented in electrical circuit diagrams.

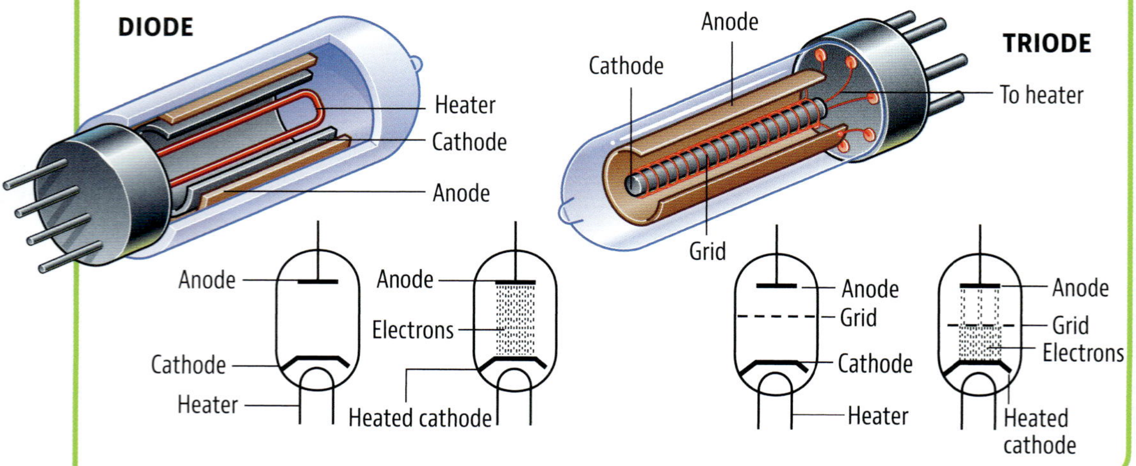

CATHODE-RAY TUBE

In early cathode-ray tube oscilloscopes, electrons were focused to form a dot on a screen. The beam was repeatedly swept from side to side and up and down to trace a graph showing measurements of, for example, a nerve signal (below).

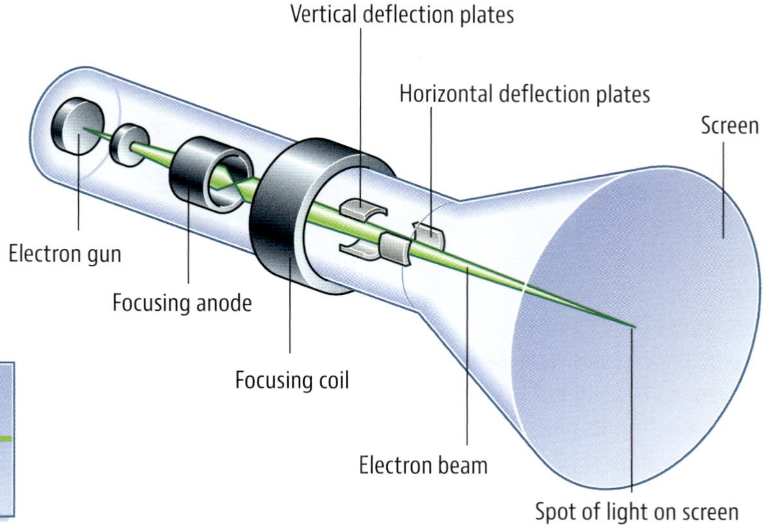

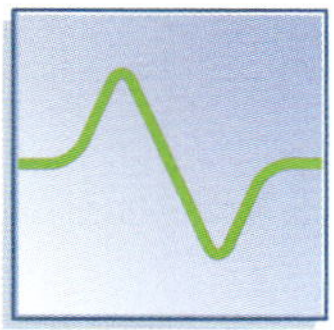

As far as current flow is concerned, a diode is a one-way device, called a rectifier. Rectifiers are important in electronics. Power is normally supplied with an alternating (reversing) current, or AC, but there is often a need to turn AC into direct current (DC). That is when a rectifier is needed.

SCIENCE WORDS

Oscilloscope: An electronic device that displays a signal as a graph on the screen of a cathode-ray tube. Now on digital displays.
Rectifier: An electronic device through which current can flow in one direction only. It converts alternating current to direct current.
Triode: An electronic device that has three electrodes. The word usually refers to vacuum tube triodes, which formerly were widely used as amplifiers.
Vacuum tube: An electronic device consisting of a glass vessel containing a partial vacuum through which electrons flow from a heated cathode.

Amplifying Currents

Currents often need to be amplified, or increased, in strength. For example, in a nondigital radio the weak current, called the signal, from the antenna must be strengthened to drive the loudspeaker. The pattern of waves in the signal must be copied exactly.

A triode can carry out the job of amplifying. The triode is a vacuum tube with a third electrode called a grid. The grid is a metal mesh surrounding the cathode. A separate small negative voltage applied to the grid holds back the electrons so that fewer pass through the grid. The small changes in the signal voltage produce large changes in the current reaching the anode.

Cathode-Ray Tube Oscilloscope

Another development of the vacuum tube became a major scientific instrument in the 20th century, and by the 1980s had become digital. In the oscilloscope, electrons from a heated cathode called an electron gun were focused by electric and magnetic fields into a narrow beam. The beam was aimed at one end of the tube, which was shaped to form a circular screen and was coated on the inside with a substance called a phosphor. When the beam struck the phosphor, it produced a glowing dot of light. The beam made a "pencil" that could be scanned across the screen extremely quickly to trace a graph showing rapidly changing signals.

Uses of Vacuum Tubes

Vacuum tubes have now been largely replaced by LCD (liguid-crystal display) flat screens for solid-state devices such as computers, radios, and TVs, but LCD screens only became possible thanks to vacuum/cathode-ray tubes, which were the technology of the day up until the 1970s and 1980s. However, some of the applications in which vacuum tubes are still used remain important. One major use of vacuum tubes is to generate x-rays. Similar screens provide some radar displays used by navigators and air-traffic controllers.

Modern oscilloscopes with LCD displays continue to be used in many scientific and industrial instruments. Today's physicians monitor the body processes of their patients with various instruments such as the electrocardiograph (for the heart) and electroencephalograph (for the brain). However, very powerful amplifiers for large currents still use vacuum tubes rather than solid-state devices, and some rock musicians think that traditional amplifiers give a better sound than those based on transistors. Some key military equipment also uses vacuum tubes because they cannot be knocked out by electromagnetic radiation, which can destroy solid-state devices.

Radar installations at airports enable air-traffic controllers to monitor and direct aircraft movements. The high-power, high-frequency waves are produced by klystrons, a type of vacuum tube. .

SCIENCE WORDS

Amplifier: A device that increases the voltage or current (usually AC) of a signal.
Cathode ray: A stream of electrons produced by the heated cathode in a vacuum tube. Cathode-ray tubes were used in all TV sets before the advent of digital flat-screen technology.
Klystron: A vacuum tube designed to generate high-power microwaves.

Radio Transmitters

High-power radio transmitters use large vacuum tubes to generate carrier waves.

LEE DE FOREST

Electronics was revolutionized when the U.S. inventor Lee De Forest invented the triode in 1906. The triode made it possible to use radio waves to broadcast sound and not just the dots and dashes of Morse code. Just four years later De Forest, who was born in 1873, made the first broadcast of an opera. While his triode tube became fundamental to all electronics, De Forest continued to innovate. Among his many other inventions were methods of recording sound on movie film and for television. He died in 1961.

These waves are continuous transmissions that are modulated (varied) by the radio signals being broadcast. The tubes operate at various wavelengths, from short wave to long wave. Other vacuum devices produce microwaves for use in radio communications and radar. Chief among them is the klystron, which is generally made of metal.

The uses of klystron range from domestic microwave ovens to continuous-beam radar systems. Klystron tubes can also be used as microwave amplifiers. The magnetron is a similar type of vacuum tube that is favored for very high-power pulsed radar systems.

USING DIODES

The main use of diodes is to convert alternating current (AC) to direct current (DC). A diode passes current in one direction only. When an alternating voltage (left, red line) is applied to it, a one-way voltage (right, blue line) is developed.

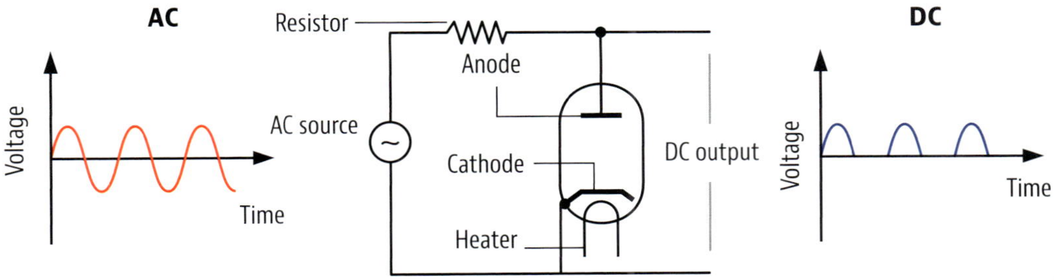

X-RAYS AND BEYOND

Physicians use x-rays to identify disease and injury in the human body, engineers use x-rays to detect flaws in industrial components, and customs officials use x-rays to discover smuggled goods and weapons hidden in baggage.

The German physicist Wilhelm Roentgen (1845–1923) noticed in 1897 that while he was generating cathode rays in a Crookes tube, a screen covered with a barium compound began to glow, even though the tube had been placed inside a cardboard box. His experiments showed that the tube was sending out a highly penetrating, short-wavelength radiation, which he called x-rays (the "x" representing "unknown"). It was found that x-rays could make images of normally invisible things, such as the bones inside living people.

X-rays are electromagnetic radiation, differing from visible light in their short wavelengths. X-rays also have a much higher energy than visible light. An x-ray is given out when an electron jumps from one energy level in the atom to a deeper one with a much lower energy. This happens when electrons bombard matter and knock other electrons out of the innermost orbits of some atoms. What happens then is that electrons from higher orbits jump into the spaces created in these lower orbits, losing energy that is given out as x-rays.

X-RAY TUBE

X-rays are created in an x-ray tube (which is similar to an electron/vacuum tube). X-rays are generated by a high-speed beam of electrons striking a tungsten plate, which has to be cooled to prevent it from melting.

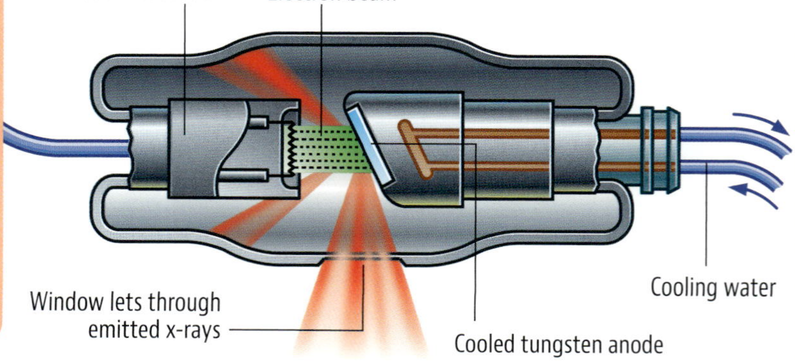

Image of part of the Milky Way taken by the Chandra X-ray Observatory Space Telescope. Many sources of x-ray light (blue dots) are stars forming. Because x-rays don't penetrate Earth's atmosphere, these would not have been visible from Earth.

Gamma Rays

Beyond the highest-energy x-rays are gamma rays, which are given off in radioactive processes. Gamma rays generally come from the nucleus, and electrons play no part in producing them. However, another important process that produces gamma rays is the annihilation of an electron with its antiparticle—called a positron.

A positron has the same mass as an electron, but in every other way it is the opposite of an electron. It has a positive charge equal to the negative charge of the electron, and it has opposite magnetic properties. Positrons are rare. When one positron is produced in a nuclear process, it soon collides with an electron, and the two disappear completely, being converted into two gamma rays.

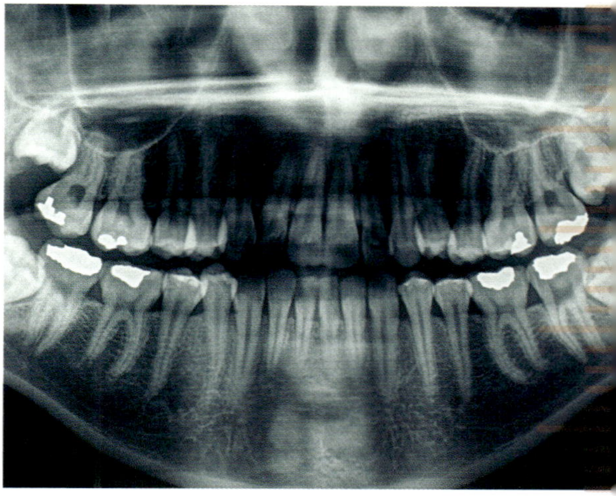

On this dental x-ray, fillings show up as bright white areas. X-rays help dentists to spot any hidden areas of infection or tooth fracture.

Rays From Space

X-rays and gamma rays are important to astronomers. They are emitted by high-energy sources such as very hot stars and the centers of some galaxies. Studying these radiations tells us more about these objects than we can learn from just the visible light they give out. Special detectors are launched on orbiting observatories because gamma rays and x-rays are absorbed by the Earth's atmosphere. Various detectors are used, some of which convert the radiation into visible light.

SCIENCE WORDS

Gamma ray: Electromagnetic radiation of high frequency (very short wavelength). While x-rays are emitted by electrons outside the nucleus, gamma rays are emitted by the nucleus itself.

X-ray: Penetrating electromagnetic radiation of very short wavelength and high energy. X-rays are produced by electrons jumping between the deepest, innermost electron shells of the atom.

SEMICONDUCTOR MATERIALS

Compact and powerful computers, radios, TV sets, and other electronics goods are solid-state devices. This means they are based on how electrons and other charge-carrying particles move through solid crystalline materials called semiconductors. The special properties of semiconductors arise from the behavior of the electrons in their atoms.

Close-up view of a silicon diode. A square crystal of silicon can be seen between the two leads.

Electrically, materials fall into three main groups: conductors, insulators, and semiconductors. The revolutionary change away from vacuum tubes was made possible by semiconductors.

In a conductor, such as a metal, more current flows if a greater voltage is applied to it. There are always huge numbers of free electrons in the metal that have been separated from their atoms. In an insulating material, such as rubber or many kinds of plastic, there are very few free electrons, and large voltages can be applied to them without large currents flowing.

Semiconducting materials, such as the elements germanium, silicon, and gallium, have very few free electrons at low temperatures, but more as the temperature increases. A temperature increase means more vibration of the atoms, and this "shakes" electrons free from some of the atoms. Shining light onto some semiconductors also "shakes" more electrons free.

The different energy levels in which electrons are arranged in atoms are known as shells. Each shell can hold only a certain number of electrons. The innermost shell holds two, and the next holds eight. The third shell can be "satisfied" with either eight or 18 electrons—it does not strongly tend to lose or gain electrons when it has either of these numbers. Atoms become more stable by losing or gaining electrons so that they have only filled shells. This explains the ways in which atoms combine with one another—they either lose or gain electrons, or share them, to achieve filled shells.

Elements that are good electrical conductors, such as metals, can easily lose their outermost electrons, which form a "sea" throughout the material and are easily set in motion when a voltage is applied.

Silicon, with 14 electrons in its atom, has the first two shells filled, but has only four electrons in the outermost shell. However,

SCIENCE WORDS

Conductor: A material or object that allows electric current (or heat) to flow through it. See also Insulator.
Insulator: A material or object that is a poor conductor of electric current (or heat).
Semiconductor: A material that has a resistance intermediate between those of an insulator and a conductor.

TRY THIS

Moving Holes

In this project we model a *p*-type semiconductor, which is germanium doped with aluminum ("doping" is the scientific term given to the process of adding a few "foreign" atoms to a semiconductor material).

What to do

Cut a piece of paper 7 in (14 cm) square and, using a ruler and black pencil, mark it out in 0.5 in (1 cm) squares. Cut two slightly bigger squares of cardboard and glue them together. Glue the paper grid onto the cardboard and use colored pencils to mark the positions of the germanium and aluminum atoms as shown in the illustration. Make the germanium atoms green and the aluminum atoms purple. Pin large-headed map tacks (all of one color) to represent the electrons around each atom. You need eight around the germanium atoms but only seven around each aluminum atom. Aluminum has one less electron than germanium. This leaves gaps in the grid (indicated by black arrows)—the technical name for such gaps in the atomic structure is "hole."

When a source of electric current is connected across a *p*-type semiconductor, it is the holes that "flow" through the material from the positive connection to the negative one. Take an electron (a tack) from a germanium atom next to the aluminum atom in the bottom left-hand corner, and plug it into the hole round the aluminum atom. Now move this tack to the next vacant hole. Then move it to the next, and so on. This is how a *p*-type semiconductor carries electric current.

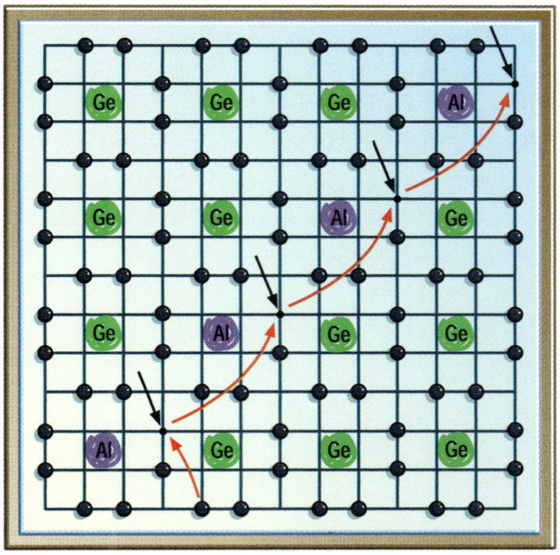

Red arrows indicate the movement of electrons

silicon atoms can share electrons with one another. In a silicon crystal, each atom has four neighbors. By sharing its electrons with these neighbors, an atom effectively has eight electrons in its third shell. At low temperatures, such shared electrons are tightly bound in place and cannot flow. However, at higher temperatures some of the electrons are sometimes jiggled out of place. If a voltage is applied, they can move and form a current. These mobile electrons are called charge carriers. This difference in electrical behavior at different temperatures means that silicon is classed as a semiconductor by scientists.

Impure Semiconductors

Adding a small proportion of atoms of a different type can increase the number of charge carriers. An example is arsenic. It has 33 electrons. Of these, 28 fill the first three shells (2 + 8 + 18), and five lie in the outermost shell. A few arsenic atoms can be fitted into the crystalline grid of silicon atoms. They share four electrons with electrons from four neighboring atoms, but one is left over. This can go wandering for large distances through the crystal. The arsenic atom left behind is now positively charged and is called an ion. Mobile electrons are negatively charged. This sort of semiconductor is called *n*-type ("*n*" for negative).

BRATTAIN, BARDEEN, AND SHOCKLEY

The transistor that revolutionized electronics was the brainchild of three researchers working at the Bell Telephone Laboratories. Walter Brattain (1902–1987) was raised on a cattle ranch. John Bardeen (1908–1991) had worked as a geophysicist before joining Bell. William Shockley (1910–1989) had directed research into antisubmarine warfare during World War II.

First, Bardeen and Brattain devised a type of transistor known as the point-contact transistor, using a germanium crystal. Then Shockley invented the junction transistor, which was to become the most widely used type of transistor. The three of them shared the 1956 Nobel Prize in Physics.

Shockley went on to head his own Shockley Transistor Corporation. Bardeen became a professor at Illinois University and in 1972 shared another Nobel Prize in Physics with Leon Cooper (1930–2024) and John Schrieffer (1931–2019). They had developed the BCS theory (named for their initials) that explained superconductivity as being the loss of electrical resistance by some materials that occurs at very low temperatures.

SCIENCE WORDS

***n*-type semiconductor:** A semiconductor material in which current consists mostly of electrons in motion.

***p*-type semiconductor:** A semiconductor material in which current consists mostly of moving holes.

Transistor: A solid-state electronics device that amplifies a small signal current or voltage and turns it into a large output current or voltage. The two major types are called junction transistors and field-effect transistors.

Aluminum is an example of another type of impurity atom. Each of its atoms has 13 electrons, with 10 in the first two shells and three in the outermost shell. If aluminum atoms are added to the silicon grid, they can share electrons with three neighbors, but each atom will be one electron short of a filled shell. This gap is called a "hole."

Occasionally a nearby electron may jump into such a hole, leaving a positively charged hole at the point from which it jumped. It is just as if the hole had jumped in the opposite direction. If a voltage is then applied across the silicon crystal, electrons will move in one direction. Those that are near holes jump into them, leaving holes behind. Other electrons jump into those holes, and so on. In this way, electrons move across the crystal in one direction, while positively charged holes appear to move in the opposite direction.

Engineers view this silicon "doped" with aluminum as being filled with mobile positive holes. There is also a small proportion of mobile electrons, freed from their parent atoms by thermal vibrations, but it is holes that dominate. Similarly, in *n*-type material, although electrons dominate, there are a few holes present. Material in which charge is mostly carried by holes is called *p*-type ("p" for positive). When a hole moves a long way, what really happens is that many electrons each hop a short distance from one atom across to a neighboring one.

Putting It All Together

Scientists had known about the existence of these semiconductors since the late-19th century. However, it was only in the 1940s that three researchers at the Bell Telephone Laboratories in New Jersey discovered how to use them in devices that would do the job of vacuum tubes. The first transistor, a tiny device, could do the job of a bulky triode:

n-TYPE AND *p*-TYPE SEMICONDUCTORS

An arsenic impurity atom in a lattice, or grid, of germanium atoms has an extra electron (left). Such electrons wander through the lattice, making it an *n*-type semiconductor. An aluminum atom has an electron missing from its outermost shell, creating a "hole" that wanders around like a positive charge (right), making it a *p*-type semiconductor. When an electric field is applied to the semiconductor **(A)**, **(D)**, electrons move toward the positive terminal **(B)**, **(C)**, but holes drift toward the negative terminal **(E)**, **(F)**.

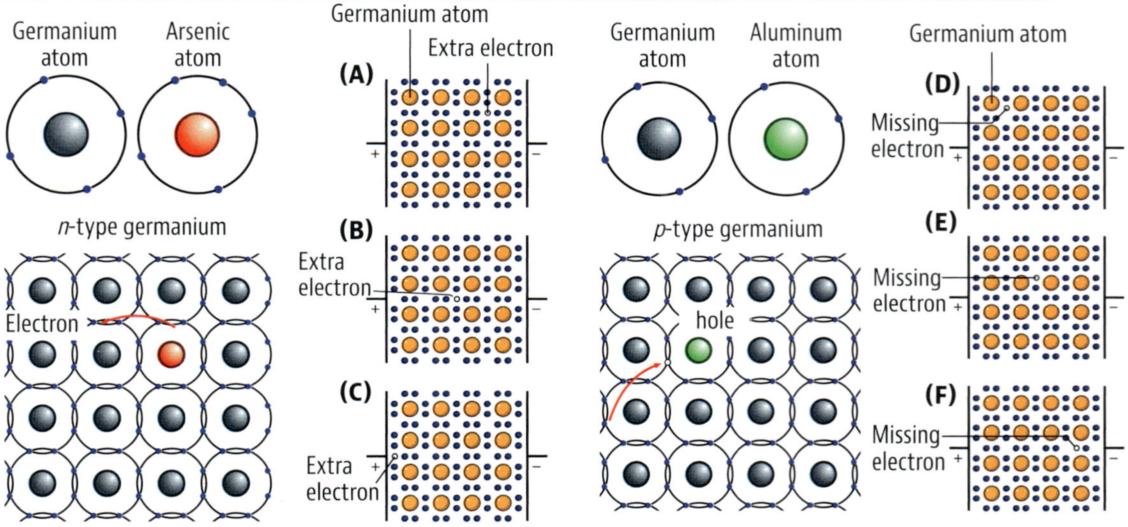

it could amplify a voltage. That is to say, if a weak varying signal voltage was applied to it, a larger voltage was generated that had the same wave pattern as the original.

Its small size was not the only advantage that the transistor (the word is an abbreviation of "transfer resistor") had over the vacuum tube. It did not need to be heated, so it used much less energy than a vacuum tube. It was also more reliable.

The first transistor design was soon improved, and within a few years commercial products containing semiconductor devices began to appear. It was not only triodes and other amplifier devices that were replaced. Vacuum-tube diodes were also replaced by semiconductor diodes. Computers, radios, and TV sets all became "transistorized."

A replica of the first transistor, developed in 1947 at the Bell Telephone Laboratories. The invention of the transistor heralded the beginning of the age of solid-state electronics.

SOLID-STATE DEVICES

Since the invention of the prototype transistor in 1947, a huge number of different types have been developed. Each type is designed to serve a certain specialized purpose. All rely on the fact that in semiconductors, electric current is carried by two sorts of charge—negative electrons and positive holes.

A solid-state diode consists of a piece of *p*-type semiconductor in close contact with a piece of *n*-type semiconductor. For example, one kind of diode is made by taking a piece of *p*-type material and adding impurity atoms to a small region to turn that part into *n*-type material. The boundary between the two types of material is called a junction. Alternatively, a small region of *p*-type material can be created within a piece of *n*-type material by adding impurity atoms of a different kind.

LED (light-emitting diode) lighting is extremely energy efficient, generating far greater light output per unit of power input than traditional incandescent bulbs.

Forward Bias

If the *p*-type material is connected to the positive terminal of a battery, and the *n*-type material is connected to the negative terminal this is called forward-biasing the diode. The battery pushes a stream of electrons into the *n*-type material. The electrons in the *n*-type material are therefore forced toward the junction. At the same time, the positive terminal attracts electrons out of the *p*-type material of the diode. This creates holes that are filled by electrons coming from deeper within the material. It is as though holes are flowing toward the junction—which just means that electrons are being pulled (in single-atom hops) away from the junction. Holes and electrons meet at the junction and

SCIENCE WORDS

Diode: An electronic device having two electrodes, which allows current to pass through in one direction only. Early diodes were vacuum tubes; modern ones use solid-state electronics.

LED: Abbreviation for light-emitting diode, a solid-state diode that gives out light when electric current passes through it.

Solid-state electronics: The science and technology of using semiconductor materials rather than vacuum tubes to make electronic devices.

neutralize each other. The positive and negative charges cancel, so that there is no charge overall at that point. As a result, current flows freely when the diode is forward-biased.

Reverse Bias

If the connections to the battery are reversed, the battery tries to draw charge carriers away from the junction. In *n*-type material, when electrons are pulled away from the junction they are not replaced. The *p*-type material on the other side of the junction cannot supply replacement electrons. As the negatively charged electrons are removed, they leave a region of positive charge behind. This attracts the electrons and eventually stops them being pulled any farther.

SOLID-STATE DIODES AND TRANSISTORS

A solid-state junction diode consists of a piece of *p*-type material and a piece of *n*-type material joined together. Current can flow only one way across the junction. In the junction transistor, a piece of one type of material, the base, is sandwiched between two pieces of the other kind. In the field-effect transistor the job of the base is done by the gate, consisting of one type of material (*p*-type is shown here) that has been diffused into a larger piece of material of the opposite type.

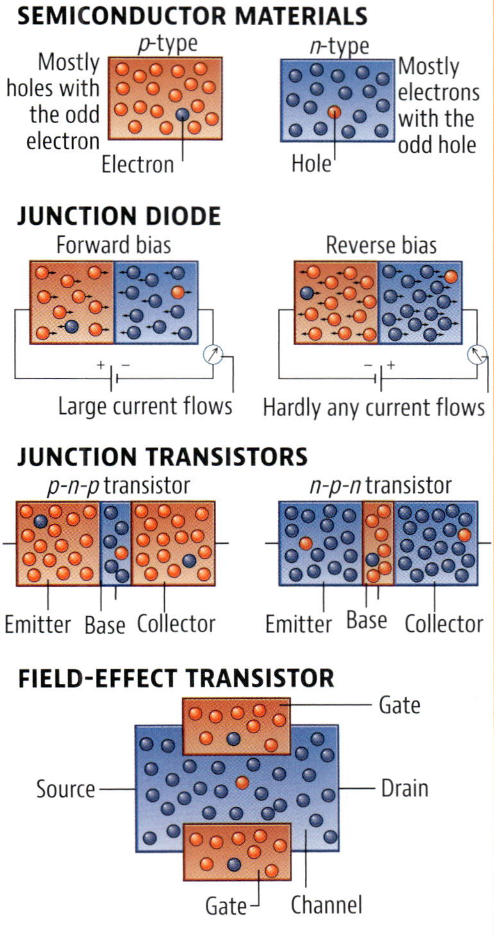

In the *p*-type material, the holes are also pulled away from the junction a little way and then stop. As a result, no current can flow through a reverse-biased diode. The fact that current can flow only one way makes the diode ideal as a rectifier in radios and many other devices.

Junction Transistors

The junction transistor consists of three types of semiconductor joined together—either *p*-type material sandwiched between two pieces of *n*-type material (called an *n-p-n* transistor), or n-type material sandwiched between p-type pieces (a *p-n-p* transistor). Or the transistor can be made from a single piece of semiconductor material containing different *p*-type and *n*-type regions. The central section is called the base, while the two ends are called the emitter and the collector.

SCIENCE WORDS

Collector: The part of a transistor that the charge carriers (electrons or holes) flow toward.
Emitter: The part of a transistor from which the charge carriers (electrons or holes) flow.
Voltage: The difference in electric potential between two points. It is measured in volts.

DIODES

Current can flow only one way through a diode. The arrowhead in the circuit symbol shows the direction of movement of positive current. A light-emitting diode (LED) is made of materials that give out light when a current flows through them.

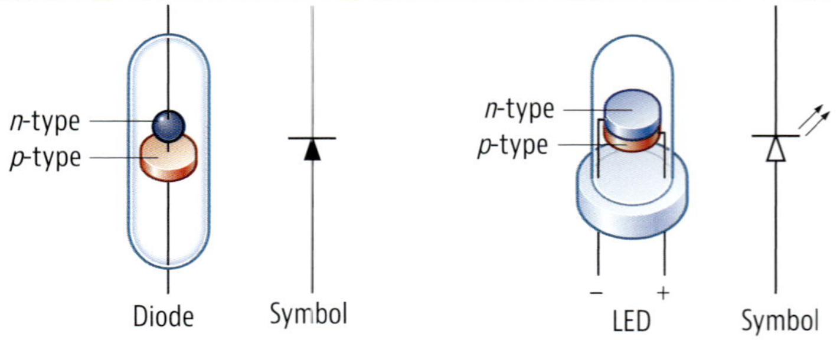

TRANSISTORS

In junction transistors (left and center) changes in base voltage control the flow of electrons from emitter to collector.

In the field-effect transistor (right) the gate voltage controls the flow of electrons from source to drain.

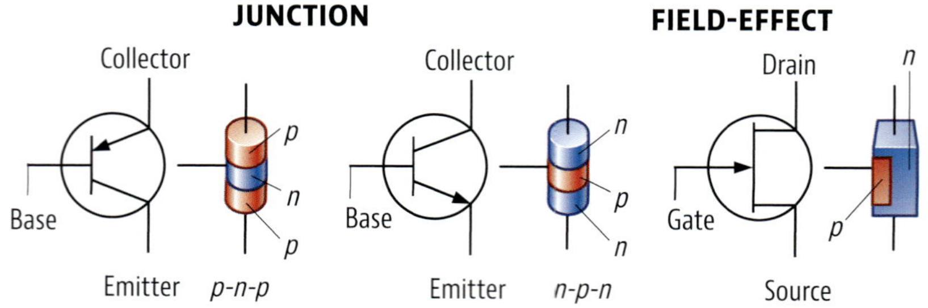

The junction transistor is like two semiconductor diodes back to back. Take an *n-p-n* transistor as an example: if a small positive voltage is applied to the *p*-type base, the emitter/base junction is forward-biased. Holes flow from the *p*-type base into the emitter, and electrons flow from the *n*-type emitter into the base.

A larger positive voltage is applied to the collector. Many of the electrons entering the base reach the junction between the base and the collector. The base is made thin to help them do this. The electrons are attracted across the junction and into the collector, and from there they pass into the external circuit.

Therefore, a small base current, consisting of electrons flowing out of the base, results in a large collector current, consisting of electrons flowing out of the collector. Thus the transistor has acted as an amplifying device.

Field-Effect Transistors

The field-effect transistor, or FET, consists of a piece of *n*-type material called the channel. There is a *p*-type region, called the gate, on each side of the channel. When a positive voltage is applied to the right end, electrons flow into the device at the left end (called the source) and leave at the right end (called the drain).

If a negative voltage is applied to the gates, the gate/channel junctions are reverse-biased. No electrons can flow into the gates because of the reverse bias. The effect of the negative voltage also extends into the channel and reduces the flow of electrons from the source to the drain. If the negative voltage of the gates is increased, the current in the channel decreases still more.

Therefore, it can be seen that in the field-effect transistor, quite small changes in the gate voltage cause large changes in the current that flows through the device. For this reason, the FET can be used to amplify signals.

TRY THIS

Radio Signals

A key part of a transistor radio is called a detector, usually a solid-state device, that detects incoming radio signals. In this project, you will produce radio signals and use a transistor set to detect them.

What to do

Bare the ends of a piece of thin insulated wire about 3 feet (1 m) long. Tune a radio to a medium-wave station. Then, move the tuning dial slightly so that you cannot hear the station, just a hissing or humming sound. Lay the wire in a loose loop over the radio. Hold one end of the wire on one terminal of the battery. Scratch the other end of the wire against the other battery terminal. What can you hear?

You should hear a crackling sound coming from the radio. The slight sparking as you scratched the battery terminal caused the current in the wire to change, and that generated weak radio signals. The signals were picked up by the radio's internal antenna and detected by its detector circuit.

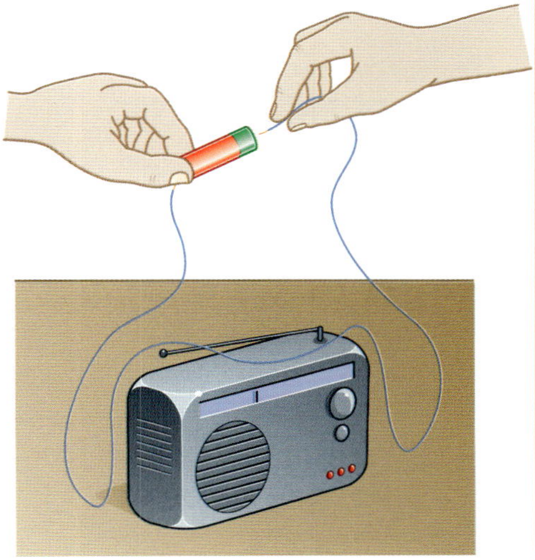

Sparks produced when you scratch the wire on the battery produce weak radio signals.

MAKING CHIPS

Microchips are literally chips of semiconductor, carrying electronic circuits. They are manufactured hundreds at a time, they are plentiful and cheap, and they provide the computing power used in "smart" devices every day.

Miniaturization of electronic devices began with the development of printed circuit boards, known as PCBs. A PCB is made from a single piece of copper foil laminated onto a plastic (nonconductive) base. The parts of the copper that are not needed are etched away with acids. What remains after this is a network of copper pathways that provide the connections between components that will be added, completing the circuit. In almost any electronic device there is at least one PCB, with a host of components mounted onto it and soldered (electrically "glued") to it. The PCB is compact, convenient to handle, and easy to manufacture in large quantities with complete accuracy—therefore making them cheap to produce.

SCIENCE WORDS

Circuit: A network of electrical components that performs some function.
Microchip: A piece of semiconductor material in which an integrated circuit has been made; it is literally chipped from a wafer in which hundreds of identical circuits are made.
Printed circuit: An electronic circuit consisting of components attached to a board or to flexible plastic and connected by lines of metal printed onto the support.
Wafer: A slice cut from a cylinder of semiconductor material in which hundreds of integrated circuits have been made. The wafer is cut up into microchips.

A printed circuit board (PCB) before the addition of components. The first PCB was made in 1936, but the technology was only used on a large scale from 1943 onward, when the U.S. Army used it to make field radios.

Integrated Circuits

A printed circuit gets its name because an essential part of its manufacture involves printing the circuit design onto the board. The design remains while the copper in other areas is etched away (see diagram, right). The same principle is used in making integrated circuits. An integrated circuit consists of various electronic components formed in a single piece of semiconductor material.

Making a Microchip

A microchip is made from a rod of very pure crystalline semiconductor. The following description applies to silicon, the most commonly used material. It is manufactured so that there is less than one unwanted impurity atom in 1 billion. But there are controlled amounts of a desired impurity, such as boron or indium, which is what

MAKING A PRINTED CIRCUIT

1 A nonconducting base metal (brown) is coated with copper and then with a photoresist (light sensitive) material (blue). **2** An opaque film carrying the transparent circuit design is placed on top and illuminated with ultraviolet light (green). The parts exposed to light are "hardened." **3** Acids remove the unexposed photoresist and the copper areas beneath. The circuit stays intact. **4** Areas of hardened photoresist are removed, leaving the copper circuit.

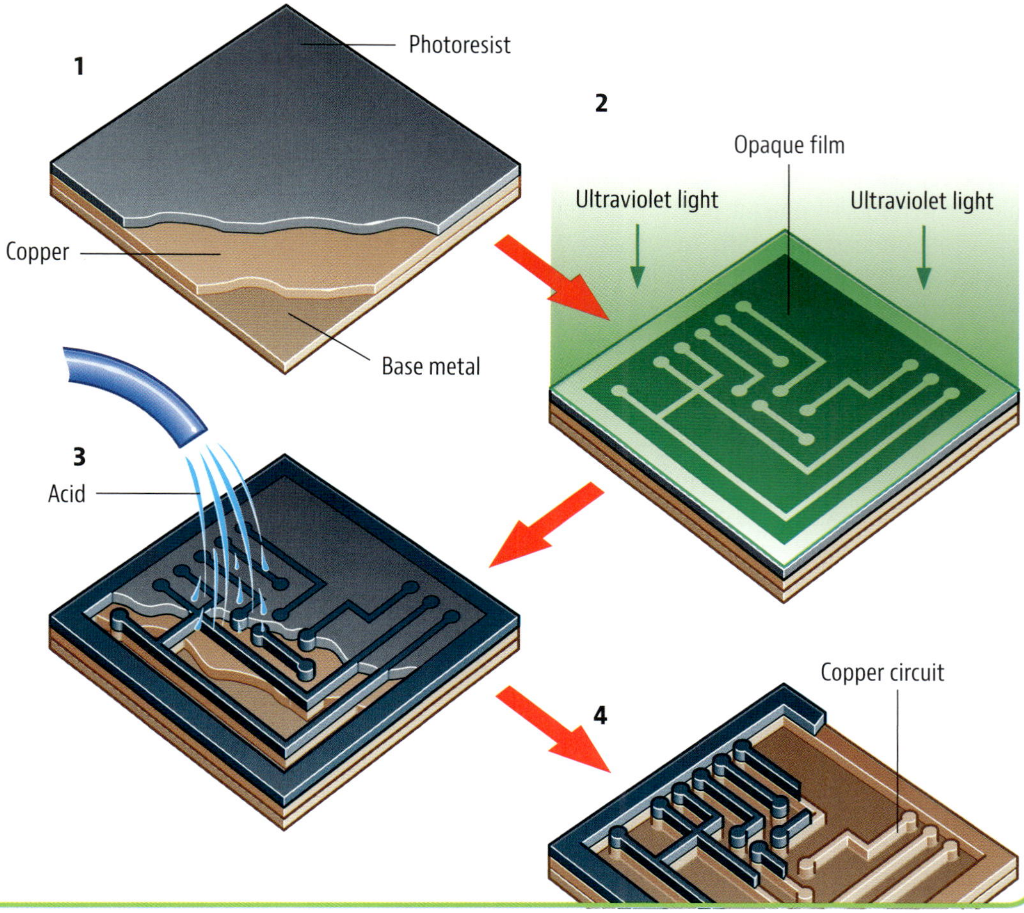

makes the silicon into *p*-type material. A thin circular disk approximately 2 inches (5 cm) across is cut from the rod and polished to form a perfectly flat wafer about $^1/_{125}$ inches (a fifth of a millimeter) thick. Hundreds of microchips can be made from a single wafer.

The *p*-type material forms the underlying substrate of the microchips, an inactive supporting layer. Areas of *n*-type material are formed in the upper part of the substrate, each one the starting point of a component such as a transistor or a diode.

When the finished component is used, a negative voltage is connected to the *p*-type material, which means it is reverse-biased in relation to the regions of *n*-type material.

This also means that no current can flow from each component into the substrate, and so they are all well insulated from one another.

The Silicon Dioxide Coating

The wafer is put into an oxidation oven where atoms of oxygen combine with the silicon in the upper layer of the wafer, about 20 millionths of an inch (about half a micrometer) thick. This is coated with a material called photoresist. Then a film called a mask is placed over the wafer. Most of the film is clear, but the circuit design appears on it as a network of opaque lines. When ultraviolet light is shined onto the mask, it passes through the clear parts and "hardens" the photoresist.

The mask is removed, and acid is applied to etch away the parts of the photoresist that were not exposed to ultraviolet light—the part representing the circuit. The acid also removes the silicon dioxide layer directly beneath, and as a result the circuit pattern is now represented by the exposed surface of *p*-type silicon.

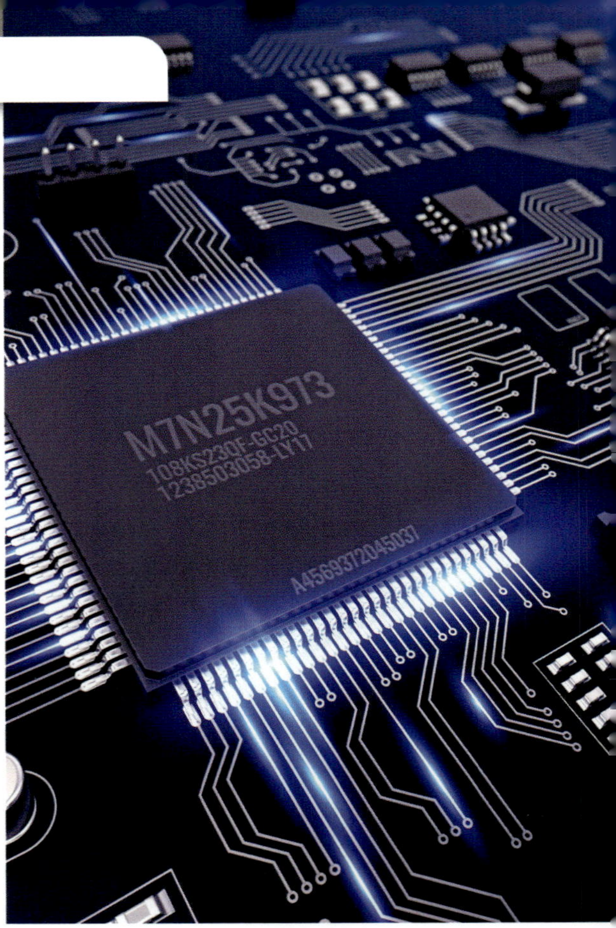

Electronic printed circuit board with processor and microchips. Advanced integrated circuits now control all the many electronic devices we use in our daily lives, from washing machines and microwave ovens to cell phones and computers.

Creating Components

The wafer now goes into an oven in which there is a vapor of an element such as phosphorus. The phosphorus atoms diffuse into the exposed areas of the silicon, forming a network of *n*-type material.

To build up components, more atoms have to be diffused into certain parts of this *n*-type network. To do this, layers of silicon dioxide and photoresist are again formed on the wafer. A new mask with a different pattern is used to print another network onto the resist, and again unexposed resist and the silicon dioxide layer beneath are removed to expose relevant areas in the surface of the previously created *n*-type areas. This time atoms such as boron or indium may be diffused in to create *p*-type regions, or further doses of phosphorus may be used if higher concentrations of electrons are needed in parts of the *n*-type regions. Finally, the components are linked with aluminum connections joined to the upper surfaces of the various *n*-type and *p*-type regions in the chip. Some microchips also use electrical connections within the semiconductor.

A microprocessor made this way can contain billions of separate transistors, diodes, capacitors, and resistors in a space of a few square millimeters. The hundreds of such circuits on the wafer are electronically tested,

MAKING A MICROCHIP

Hundreds of identical microcircuits are made simultaneously in several layers within a single wafer of semiconductor. This sequence shows the production of part of a single circuit, starting with a piece of *p*-type silicon.

1

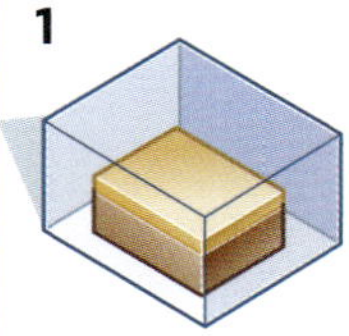

A p-type silicon wafer is baked in an oxidation oven to 1,830°F (1,000°C), which produces a thin protective layer of silicon dioxide on the wafer's surface.

2

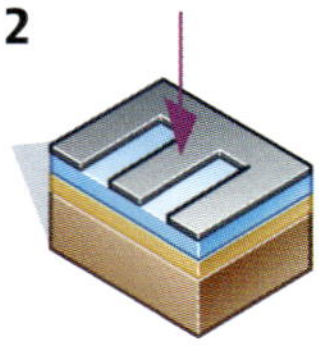

A photoresist coating is applied to the silicon dioxide layer, and a film mask of the circuit design is laid over it. Ultraviolet light is shined through the circuit mask.

3

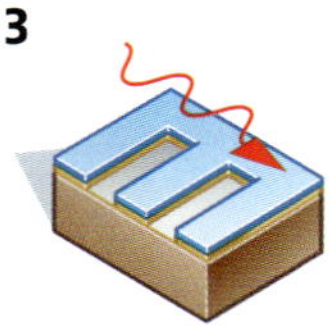

The mask is removed, and the wafer is treated with a developer that removes the areas of photoresist not exposed to ultraviolet light. These are the areas that represent the circuit components.

4

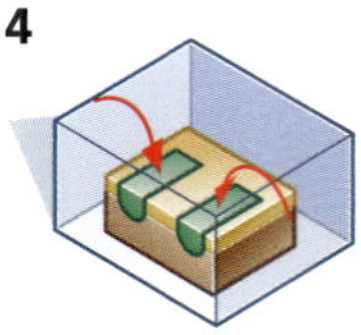

Hydrofluoric acid etches away the silicon dioxide where photoresist has been removed. The rest of the photoresist is then removed. The wafer is exposed to phosphorus vapor in an oven. Phosphorus atoms diffuse into the silicon, creating n-type regions where silicon dioxide is removed.

5

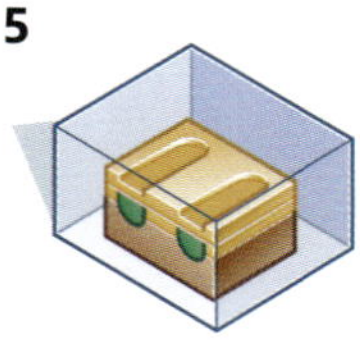

The wafer is returned to the oxidation oven, where another layer of silicon dioxide is added in preparation for further stages of etching.

6

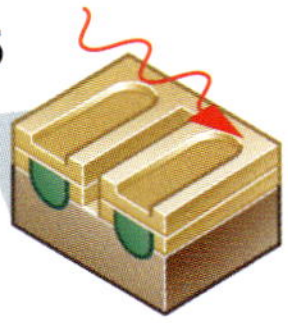

The wafer is given another coating of photoresist, and further masking and etching stages are carried out.

7

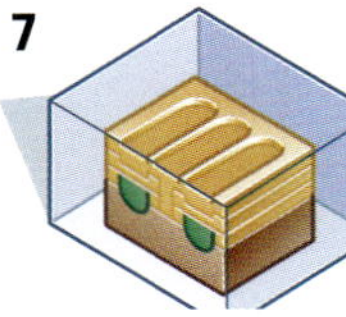

The wafer is returned to the oxidation oven to add another layer of silicon dioxide.

8

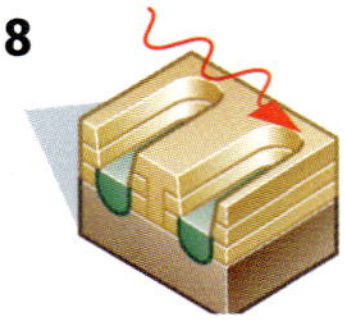

The next stages of masking and etching make the channels previously created deeper and narrower.

9

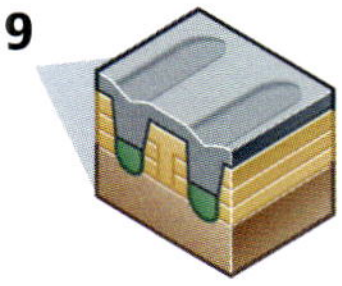

The wafer is placed in a vacuum chamber and exposed to aluminum vapor. A layer of aluminum is added to the wafer.

10

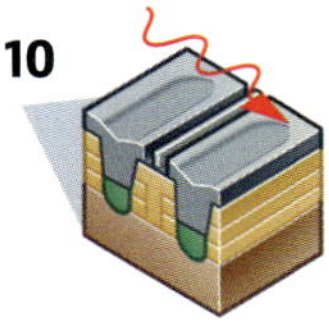

The wafer goes through a final series of masking and etching stages that form the electrical connections for the complete circuit.

defective ones are marked, and the wafer is broken up and the faulty circuits discarded.

Wires are connected at various points to the good chips, and they are mounted in containers, which may be round but are more often rectangular. In most applications, the microchip capsules are mounted on a printed circuit board.

ELECTRONIC MEMORIES

At any one time a computer has to hold a huge amount of stored data. The physical devices that hold this information make up its "memory." There are many different kinds of memory. The amount of information that computer memory is capable of holding is expressed in bits and bytes.

Computers work with what is called a binary code, in which all numbers are represented as a sequence of zeros and ones instead of the ten digits from 0 to 9 that are ordinarily used. A single binary digit, 0 or 1, represents one bit. A string of eight bits is called a byte and can represent any of 256 different numbers (from 0 to 255). In some contexts the computer treats them as numbers, and in others it treats them as language characters. The 26 letters of the alphabet, both lower-case (small) and upper case (large), punctuation characters, and some other symbols can each be represented by a different byte. The maximum size of a memory is usually expressed in bytes, kilobytes (1 kB = 1,024 bytes), megabytes (1 MB = 1,048,576 bytes), gigabytes (1 GB = 1,073,741,824 bytes), and terabytes (1 TB = 1,099,511,627,776 bytes).

The computer's short-term memory contains data that it is directly working with, including intermediate results in the course of a calculation. This type of information has to be available immediately, so it is stored in RAM (random-access memory), where any piece of information can be accessed equally quickly. RAM comes in special microchips called RAM chips. The retrieval time from a chip for a single piece of information—known as the access time—can be as little as 5 nanoseconds (billionths of a second).

SCIENCE WORDS

Bit: A unit of information representing a choice between two possibilities. In the binary system, in which all numbers are written with the numerals 0 and 1 only, a single 0 or 1 represents one bit.
Byte (B): A sequence of bits, usually 8. In computing, this means a sequence of 8 digits—that is, 0's or 1's.
Gigabyte (GB): In common usage 1,000 megabytes.
RAM: Abbreviation for random-access memory, memory storage in a computer in which any item of information can be retrieved for processing equally quickly.
ROM: Abbreviation for read-only memory: in a computer, memory where the contents are not altered by operation.

Types of RAM

There are two main types of RAM. Dynamic RAM, or DRAM (pronounced "dee-ram"), has to be "refreshed" thousands of times per second—that is, the same data has to be written into it continually. This slows down the rate at which the data can be read or altered, but DRAM is cheap and so it is the most widely used kind of RAM. Unlike DRAM, static

A disk array of hard drives in a data center. This and similar modes of storage, combined with a server, create a network through which multiple users can access files.

RAM, or SRAM (pronounced "ess-ram"), does not need to be refreshed. It is more expensive than DRAM, but faster. Both SRAM and DRAM lose their data if the power supply fails. While a computer is working, the information that is stored in RAM is constantly changing. There is other information that the computer needs, and which does not need to change. It is stored in ROM (read-only memory), which is also random-access memory, but contains data that is not altered. It may contain "boot-up" information, which is used when the computer is first turned on, and other information that is used in the normal running of the computer.

Mass Storage

Data produced by the computer needs to be stored somewhere. Hard disk drives (HDDs) provide high storage capacity. A drive contains one or more rigid disks, or platters, mounted on a spindle, and spinning very fast. Magnetic read/write heads, moving as a unit, swing back and

HARD DISK DRIVE

Backup memory that needs to be accessed rapidly is usually kept on hard disks. A large memory store consists of a stack of hard disks, mounted on the same spindle and accessed simultaneously. Disks can rotate at speeds of 100 mph (160 km/h) or more. Data is stored as magnetic fields in areas of the disk as small as 40 millionths of an inch (1 micrometer) across. Read/write heads mounted on the ends of arms swing in and out across each disk as required, "flying" a few millionths of an inch (fraction of a micrometer) above the surface, kept clear of it by a cushion of air. The carriage moves the read/write head in and out.

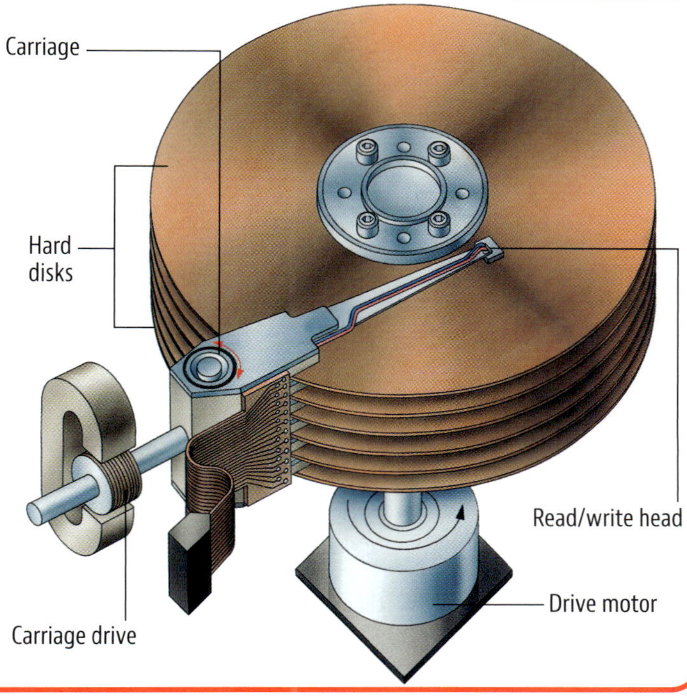

forth over each disk, reading the top and bottom sides. The magnetic coating is iron oxide or some other magnetic material. It consists of tiny regions called magnetic domains. Each domain has its own magnetic field. The head's magnetic field rotates the domain fields so that they point one way to represent a 0 or the opposite way to represent a 1. Later the read/write head can read back the data it has recorded by sensing the directions of each of the domain fields. A single piece of information can be retrieved from the disk in about a hundredth of a second.

Even larger storage capacity is often provided by magnetic tape. The tape stores information in basically the same way as magnetic disks, audiotapes, and videotapes. To access data on a tape, the tape has to be played all the way through to the appropriate point. If a program is to use the data, it must first be copied to a faster medium such as a magnetic disk. For this reason, tape is often used to make secure copies of information such as bank customers' past transactions.

CDs, DVDs, Memory Sticks

A CD-ROM is a compact disk working in essentially the same way as music CDs. Its name is an abbreviation of "compact disk—read-only memory." It contains information that a computer can read, but in its standard form the information on it cannot be altered. A CD-ROM consists of an aluminum disk covered with a protective coating of plastic. Tiny pits are etched into the underside of the disk along a track that spirals outward from the center. In the CD-ROM drive, a read head bounces a laser beam off the disk as it rotates. At any point along the track there may be a pit, in which case the laser beam is not reflected to the detector, or there may be a "flat," where a pit is missing, in which case the light is reflected to the detector. The sequence of pits and flats causes a fluctuating signal in the detector, which is converted and sent to the computer. A CD-ROM can store 670 MB of data, which is small by today's standards, and it is much slower than a hard disk. However,

EARLY TAPE DRIVE

In the late 1950s and early 1960s, this type of tape drive was used in mainframe computers. Backup data stores were kept on magnetic tape. The tape was driven past a read/write head by the drive capstan wheel. The read/write head applied a strong magnetic field that aligned tiny magnetic regions in the metal tape coating to represent bits (units of information). Today's tape drives are more like CD/DVD players.

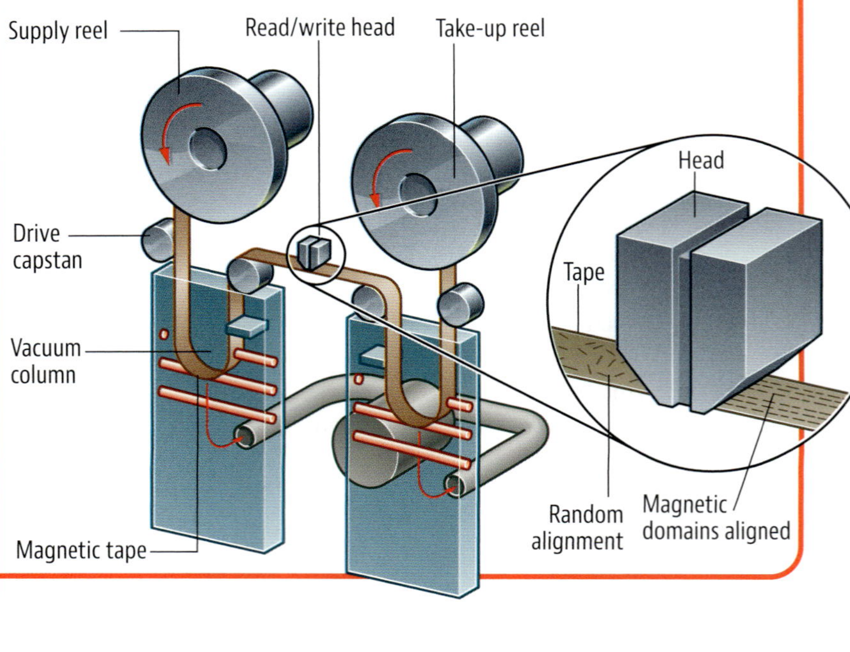

EARLY CD DRIVE

In the 1980s, CD players had more complicated optics (below) than they do now. Laser light scanned the pitted undersurface of a CD. Variations in the reflected beam were picked up by a photodiode detector and turned into sound and images or used as computer data.

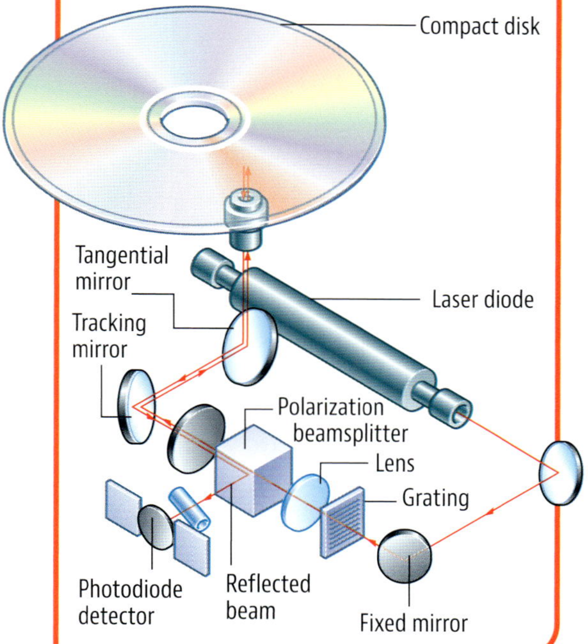

it is cheap and can safely be removed from a drive and handled.

The DVD (digital versatile disk) is a compact disk with a much higher capacity, measured in gigabytes, and a higher access speed. Older desktop and laptop computers incorporated drives that could write to special CDs and DVDs. Typically, a DVD can hold seven times more data (4.7 GB) than a CD, though some are capable of storing over 17 GB. DVDs were favored for viewing movies before the advent of online internet streaming services. The extra facilities that a DVD computerized medium offered included quick repeats or skips and other easily accessible extra material, such as background information about the stars, or alternative "takes" from the movie.

Memory sticks are storage devices that plug into a computer's USB port. These are ideal for copying data such as files from a digital camera onto a personal computer. They are light and portable, and can store up to 2 TB of data. Memory sticks use flash memory, also used in cell phone memory cards and solid-state drives (SSDs), which replaced the hard drive in many computers. Flash memory uses special FET transistors that store information even when their power is switched off. Because there are no moving parts, retrieving data is 100 times faster than a hard disk.

In the Cloud

Cloud computing is a system where computer files—emails, pictures, music, documents—are not stored on a device. Instead the information is stored in "the cloud." This is not literally in the air, but in huge data centers. One person's files will be stored in many centers all over the world. Together, these centers form a "cloud" of computers that can be accessed via the Internet from anywhere in the world by a wide range of devices.

SCIENCE WORDS

SSD: A data storage device in a computer. It uses flash memory to store data.
Flash memory: A data storage device for computers and electronics, most often used in digital cameras and video games.
DVD: A type of optical disk that can store much more information, and retrieve it at a higher speed, than an ordinary CD.
Hard disk: A rigid magnetic disk used to store large quantities of computer data.

GLOSSARY

Alternating current (AC) Electric current that flows first in one direction, then in the other, alternating many times each second. AC is used for domestic electricity supply and many other electrical applications.

Ammeter An instrument for measuring electric current.

ampere (A) The SI unit of electric current. A current of 1 ampere (often abbreviated to "amp") is equal to a flow of 1 coulomb per second.

Anion An ion with a positive charge.

Anode A positive electrical terminal on a device such as a electron tube. Electrons flow into the device through the anode.

Capacitance Also called capacity, the ratio of the stored electric charge on an electrical device or other electrical object to the voltage applied to it.

Cathode A negative electrical terminal on a device such as an electron tube. Electrons flow into or out of the device through the cathode.

Cathode ray A stream of electrons produced by the heated cathode in a vacuum tube; cathode-ray tubes were used in all TV sets before the advent of flat-screen technology.

Cation An ion with a negative charge.

Charge A property of some subatomic particles and some objects that makes them exert forces on one another. There can be negative charge and positive charge.

Collector The part of a transistor toward which the charge carriers (electrons or holes) flow.

Condenser Also called a capacitor, a device that can store electric charge.

Conductor A material or object that allows electric current (or heat) to flow through it.

Conservation of charge In an isolated electrical system, the overall electric charge remains constant.

coulomb (C) The SI unit of electric charge.

Dielectric A nonconducting material inserted between the plates of a condenser.

Diode An electronic device, having two electrodes, that allows current to pass through in one direction only. Early diodes were vacuum tubes; modern ones use solid-state electronics.

Direct current (DC) Electric current that flows in one direction all the time, though it may vary in strength.

Electric field A region around a charged object in which another charge experiences a force.

Electrolysis The decomposition of an electrolyte by electric current, using two electrodes, the anode and cathode.

Electrolyte A liquid (a solution or a molten substance) that conducts electricity, as in a battery or in electrolysis.

Electromagnetic radiation Energy that is transmitted through space or a material medium in the form of electromagnetic waves.

Electron A subatomic particle, found in every atom, which carries negative charge. Most currents consist of electrons in motion.

Electrostatic induction The production of an electric charge in an uncharged object by a nearby charged object.

Emitter The part of a transistor from which the charge carriers (electrons or holes) flow.

Gamma ray Electromagnetic radiation of high frequency (very short wavelength). While x-rays are emitted by electrons outside the nucleus, gamma rays are emitted by the nucleus itself.

Insulator A material or object that is a poor conductor of electric current (or heat).

Ion An electrically charged atom or group of atoms that has either lost one or more of its electrons (to form a positive ion) or gained one or more electrons (to form a negative ion).

Ionic bond A type of chemical bond formed between ions of opposite charge.

Magnetic field The pattern of magnetic influence around an object.

Mirror oscilloscope An electronic device that displays a signal as a graph on the screen.

***n*-type semiconductor** A semiconductor material in which current consists mostly of electrons in motion.

ohm (Ω) The SI unit of resistance.

Parallel Two electrical components in a circuit are connected in parallel if the current divides to pass through them separately.

Potential difference Also called voltage, the difference in electric potential between two points. It is measured in volts. The higher the potential difference, the greater the force tending to move charges between the points.

***p*-type semiconductor** A semiconductor material in which current consists mostly of moving holes.

Quantum theory Theory based on the idea that light is emitted in separate packets, or quanta (also known as photons).

Rectifier An electronic device through which current can flow in one direction only. It converts alternating current to direct current.

Resistance A measure of how a material or a component resists the passage of electric current through it. The higher the resistance, the less current will pass when a given potential difference is applied across it.

Resistor An electrical component with a known resistance, used to regulate current and voltage in a circuit.

Semiconductor A material that has a resistance intermediate between those of an insulator and a conductor.

Series Two electrical components in a circuit are connected in series if the same current passes through both of them in turn.

Static electricity An electric charge on an object that has lost or gained electrons.

Transformer A device that increases or decreases the voltage of alternating current.

Transistor A solid-state electronics device that amplifies a small signal current or voltage and turns it into a large output current or voltage. The types are junction transistors and field-effect transistors.

Triode An electronic device that has three electrodes. The word usually refers to vacuum tube triodes, which formerly were widely used as amplifiers.

Vacuum tube An electronic device consisting of a glass vessel containing a partial vacuum through which electrons flow from a heated cathode.

volt (V) The SI unit of potential difference.

Voltage The difference in electric potential between two points. It is measured in volts.

x-ray Penetrating electromagnetic radiation of very short wavelength and high energy.

FURTHER RESEARCH

Books

Al-Khalili, Jim. *The Physics Book* (DK Big Ideas). New York, NY. Dorling Kindersley, 2023.

Holzner, Steve. *Physics for Dummies*. Hoboken, NJ. Wiley, 2021.

Lowe, Doug. *Electronics All-In-One for Dummies 3rd Edition.* Hoboken, NJ: Wiley, 2022.

Web Sites

explainthatStuff
https://www.explainthatstuff.com/electricity.html
Simple introduction to electricity and electromagnetism, including a timeline and further reading.

Kiddle
https://kids.kiddle.co/Electricity
Facts and information about electricity, how it is made, and how it works.

Science Buddies
https://www.sciencebuddies.org/stem-activities/subjects/physics/eighth-grade
Physics activities for Kids.

INDEX

A
alkalis 23
alpha particles 4, 5
alternating current (AC) 38, 39
  changing voltages 39
  converting to DC 43
aluminum 20, 48
ammeters 31, 32, 33
amperes 29, 30
amplifier 42
anions 20, 21
anodes 4, 24, 40
atomic electricity 8, 9
atoms 4, 5
  charged 20–1
  try this 5
attraction 10, 11, 12, 13
  try this 13

B
Bardeen, John 48
batteries 28, 29, 38
binary code 58
bit 58
Bohr, Niels 7
Brattain, Walter 48
byte 58

C
capacitance (capacity) 33
capacitors see condensers
carrier waves 43
cathodes 4, 24, 40
cathode rays 5, 42
cations 20, 21
CDs (compact disks) 61
CD-ROMs 60, 61
charge(s) 8
  conservation of 12, 13
  detecting 14, 15, 16, 17
  distribution of 17
  moving 26, 27
  negative/positive 9, 10, 11
  producing 8–9
  stored 32, 33
  storing 19, 33
  testing 15
charging
  by contact 14
  by induction 14, 15
chloride 20, 23
chlorine 20
cloud, the 61
circuit 54
cloud, the 61
collector 52, 53
compact disks (CDs) 61
condensers (capacitors) 18, 19, 32, 33
  charging 32
conductors 16, 17, 28, 46, 46
  testing for 16
  try this 27
conservation of charge 12, 13
Cooper, Leon 48
Coulomb, Charles 12
Coulomb's law 12, 13
coulomb 12, 13
Crookes, William 4, 5, 6
Crookes tube 4, 6, 7
crystals 20,21
Curie, Pierre and Marie 14
current(s), electric 26, 27, 30, 31, 32
  amplifying 41
  flow of 28, 29, 30, 31
see also alternating current; direct current

D
De Forest, Lee 43
dielectric 18, 19, 33
diodes 40, 43, 50
  silicon diodes 46
  see also solid-state diodes
direct current (DC) 38, 39
  converting from AC 43
DVD (digital versatile disk) 61

E
Edison, Thomas 35
electric fields 18, 19
  point charge fields 18
  shapes of 19
electricity
  atomic 8, 9
  frictional 8
  static see static electricity
electric light bulbs 34, 35, 36, 37
  try this 35
electric locomotives 28, 29
electrocardiograph 42
electrodes see anodes; cathodes
electroencephalograph 42
electrolysis 22, 23, 24, 25
electrolyte 22, 23, 25
electromagnetic radiation 7
electromotive force (e.m.f.) 30
electrons 4, 5, 6, 7
  moving 26
  pressure of 7
electroplating 25
electroscope 16
  gold-leaf 14, 15 16
electrostatic induction 16
  try this 15
emitter 52, 53
energy 36, 37

F
Faraday, Michael 17
Faraday's ice-pail 17
friction 9, 10, 13

G
gallium 46
galvanometers 32, 33
gamma rays 44, 45
gases, producing from water 24, 25
germanium 46
gigabytes 58, 60
gold-leaf electroscope 14, 15, 16

H
halite 21
hard disks drives 58, 59 60, 61
hydrogen 20

I
incandescence 36
induction 17
insulators 16, 17, 28, 46
  try this 27
integrated circuits 54
ionic bond 22
ionic crystal 21
ionosphere 21
ions 21, 27
  formation 21
  in solution 22, 23, 24, 25
  negative (cations) 20, 21
  positive (anions) 20, 21
  producing metals/gases 24–5
  try this 23, 24

K
klystron 42, 42, 43

L
LED (light-emitting diode) 50, 50, 52
Leyden jar 18, 19
lightning 20, 21

M
magnetic field 33
magnetic tape drive 60
magnetron 43
memories, electronic 58, 59, 60, 61
memory sticks 61
metals, producing 24
microchips 54, 55, 56, 57
Millikan, Robert 6
molecules, polar 22

N
nonconductors see insulators
nucleus 8

O
Ohm, Georg 34
ohms 30, 36
Ohm's law 34, 35
Onnes, Heike Kamerlingh 37
orbitals 4
oscilloscope 41, 42

P
parallel 30, 31
  resistors in 35, 36
  try this 31
phosphor 42
photoresist 56, 57
point charge fields 18
positron 44, 45
potatoes and electric currents, try this 23
potential difference (voltage) 28, 29, 30, 31, 52
  changing 39
power
  consumption 36, 37
  resistance and 34, 35, 36, 37
power plants 29, 39
printed circuits 54, 55, 56

Q
quantum theory 7

R
radar 42, 43
radio signals, try this 53
RAM (random-access memory) 58, 59
rectifiers 40, 41
repulsion 10, 11, 12, 13
  try this 13
resistance 26, 29, 30, 31, 35, 36
  and power 34, 35, 36, 37
  combining resistances 36
  and heating 36, 37
resistive heating 37
resistors 30, 31, 35, 36
  in series/parallel 35, 36
Roentgen, Wilhelm 44
ROM (read-only memory) 58, 59
Rutherford, Ernest 4, 5, 7

S
salt 20, 21, 22, 23
Schrieffer, John 48
semiconductors 45, 46, 47, 48, 49
  impurity semiconductors 47, 48
  n-type 48, 49, 51
  p-type 47, 48, 49, 51
  try this 47
series 30, 31
  resistors in 35, 36
  try this 31
shells 46, 47
Shockley, William 48)
silicon diodes 46
sodium 20, 23
sodium chloride 20, 21
solid state devices 46, 50, 51, 52, 53
solid-state diodes 50, 51, 52
solid state electronics 50
space, rays from 45
static electricity 8, 9, 10, 27
  positive/negative 9, 10, 11
  try this 9
superconductivity 36, 37, 48
Swan, Joseph 35

T
Thales of Miletus 8
Thomson, J. J. 5, 6
transformer 38, 39
transistors 48, 49, 51
  field-effect (FET) 51, 52), 53
  junction 5), 52. 53
  point-contact 48
triodes 40, 41, 43

U
ultraviolet radiation 20, 21

V
vacuum tubes 40, 41, 42, 43
Van de Graaff generator 10
voltage (potential difference) 28, 29, 31, 52
  changing 39
volts 29

W
wafer 54, 57
water 22
  gases from 24, 25

X
x-rays 44 45
  tube 44